MICHIGAN STATE UNIVERSITY

AMERICAN POLITICAL
ECONOMY

JOHN D. MOLLOY

Top cover photo: "U.S. Federal Reserve Building," by Mark Burnett/Stock Boston.

Bottom cover photo: "U.S. Capitol Building," by Anthony Edgeworth/The Stock Market.

Copyright © 1999 by John D. Molloy.
All rights reserved.

Permission in writing must be obtained from the publisher before any part of this work may be reproduced or transmitted in any form or by any means, electronic or mechanical, including photocopying and recording, or by any information storage or retrieval system.

Printed in the United States of America

10 9 8 7 6 5 4 3 2 1

This manuscript was supplied camera-ready by the author(s).

Please visit our website at www.pearsoncustom.com

ISBN 0–536–60005–8

BA 990504

PEARSON CUSTOM PUBLISHING
160 Gould Street/Needham Heights, MA 02494
A Pearson Education Company

PREFACE

Why write a textbook called *American Political Economy?*

What is American Political Economy, and how does it differ from American Government or Economics? These are logical questions which may come to mind as students begin their journey through these pages.

Textbook authors often believe that a new approach is needed. They may feel that, although there are plenty of books in the marketplace, none of them seems to "fit" their course objectives. This is particularly true in *interdisciplinary* studies.

This book is based on a number of assumptions, including the following:

1. Introductory, undergraduate courses should be among the most interesting on campus. It is unnecessary to "clutter them up" by attempting to teach beginning students everything they need to know for the Ph.D. degree.

2. Too many college texts are written by professors to impress their colleagues or to earn tenure, rather than to *communicate effectively with students.* Although the scholarly vocabulary, rich in jargon, may delight the academic elite, it is not the language of youth or, for that matter, people in the "real world."

This book is intended to be *student-centered*, rather than based primarily on abstract theories. What interests scholars is not the primary focus. What can and should be read and absorbed by students is the criterion for selection of topics in this work.

Prior to joining the faculty at Michigan State University, I spent 15 years on the editorial staff of *The Cincinnati Enquirer*. Later, I spent many summer "vacations" there as the newspaper industry converted from "hot type" to a "cold type" (computer) technology.

As a reporter and editor, I developed a strong distaste for long, involved sentences and interminable paragraphs. Unfortunately, these are common in academic writing. Dick Havlin, news editor of *The Enquirer* in the mid-1960s, often spoke of a "fog index." This was his way of describing the writing style of those who forgot that "simple declarative sentences are best." This view was reinforced by several outstanding professors at Northwestern University's School of Journalism, where I was a visiting professor during the 1980-81 academic year.

3. Nearly all college students should be able to grasp fundamental concepts of American Political Economy if explanations are made in familiar language and examples are meaningful to them. We are not, however, trying to "dummy down" this book to the level of the least able or indolent student. It is expected that students will "stretch" their vocabularies in the college years. It is hoped that they will learn mental discipline and modes of systematic thinking while rejecting bumper sticker solutions to complex problems of public policy.

But what is American Political Economy? It is, as I see it, a study of the exercise of power and authority in both government and in the "market." It involves interaction between political, business and labor leaders, as well as other groups.

This book does not take the approach so common in introductory Political Science or Economics courses. It is designed specifically to examine problems from an *interdisciplinary viewpoint*. Political scientists and economists are taught to think as their disciplines demand. This is, of course, quite artificial and can lead to "tunnel vision." Some scholars forget that universities offered "Moral Philosophy" and "Political Economy" long before "Political Science" and "Economics" became separate fields of study.

This book is organized in numbered "articles" rather than by traditional chapters. It should be easy for students to pick up, concentrate on their reading, then put down for a "study break" without losing continuity.

Major subjects covered include the following:
- Comparative systems, theories behind them and how they actually work.
- Growth and development of the market economy.
- Technological change and its impact on American economic, political and social life.
- The development of modern business corporations and labor unions, as well as the conflicts between them.

The *economic role of government* is at the heart of this work. Considerable attention is paid to debates over the "proper" regulatory and promotional roles of the state.

Today's undergraduate students, particularly those coming from less than outstanding high schools, often lack basic knowledge of U.S. history. To remedy this, considerable attention is given to events such as The Great Depression, World War II and the development of the Cold War. Their economic, political, social and psychological impacts remain with us on the eve of the 21st Century.

It is fitting, as the presidential campaign of 2000 approaches, that much of this book deals with Chief Executives. We have chronicled their political and economic struggles with Congress, business, labor and a variety of other interest groups.

The Democrats' 20-year monopoly on the White House during the "New Deal" and "Fair Deal" had a profound effect on the way Americans think and argue about "appropriate" government economic policies. The legacies of Franklin D. Roosevelt and Harry S. Truman, as well as economist John Maynard Keynes, endure.

Should government provide jobs? If so, what kinds and how many? How shall they be financed? These questions did not end with Roosevelt and Truman. They are on today's agenda.

The legacies of Lyndon B. Johnson - the War on Poverty and the War in Vietnam - also remain with us. How much "welfare state" do we want? What can be done, if anything, when government badly miscalculates future costs of "entitlements?" Indeed, are we *"entitled"* to all the things that government provides? Should there be a *means test* for Social Security and Medicare benefits? Should *all* students be eligible for aid, or only *some* - who have demonstrated an interest in learning? Can government take "goods" away from the voting public once they have been given to it?

Are we headed for a "generational war" when the "Baby Boomers" retire? Can an increasingly smaller number of productive workers pay generous Social Security and Medicare benefits to a future "army" of retirees? These are just a few of the questions we will explore.

Among "hot" topics of debate during the 1996 presidential campaign were deficit spending, unbalanced budgets, national debt, poverty and welfare reform. How can we deal with these issues while promoting a growing and stable economy? How do we combat unemployment and prevent inflation at the same time?

Some consideration is given to the roots of the current debate over monetary and fiscal policy, Keynesian "demand-side" and Reagan "supply-side" solutions to economic problems. We survey the Reagan-Bush and Clinton Administrations and their varying philosophies and policies. Taxes and tax reform are explored to illustrate one of this book's most important points: *What is good politics may be bad economics. What is good economics may be bad politics!*

ACKNOWLEDGMENTS

Every author, in "building" a book, goes into debt to his friends and colleagues. A brief acknowledgment in the preface is appropriate, if not entirely adequate. I am indebted to several associates in the Center for Integrative Studies/Social Science at Michigan State University for their help in the production of this book. Interim Social Science Dean/CIS Director *Phil Smith* and Dr. *Bruce Bigelow* arranged secretarial assistance.

Jean Robinson of our CIS staff was most helpful with computer production and reading of the manuscript. She cheerfully and expertly has helped take this book through several drafts.

Teaching assistant *Chris Sprecher,* a doctoral candidate in Political Science, read the entire manuscript, and I value his constructive comments.

Several faculty members have stimulated my interest in interdisciplinary studies over the years. Among them are *Douglas Dunham*, chairman and professor emeritus of Social Science at Michigan State. Also, MSU Professor Emeritus *Albert E. Levak* and the late Professor *Alfred Kuhn*, a gifted teacher in the University of Cincinnati Department of Economics.

Dr. Kuhn published *Unified Social Science,* which reflected his concern that undergraduate students be taught to think first as policy problem solvers rather than as political scientists or economists.

Last, but far from least, I want to thank my wife *Carol* for her help and encouragement during the preparation of this book.

– John Molloy

DEDICATION

To Tommy and Danny North, grandchildren extraordinaire!

I. INTRODUCTION

Virtually any newspaper reader or reasonably well-informed student is aware that the U.S. has many economic problems. News stories constantly stress such troubles as the decline in the nation's competitive ability and the educational skills of its workers. Some Americans fear the threat of renewed inflation, consequent higher prices and increases in the cost of living. Although recent unemployment figures are relatively good, politicians and others argue about government bookkeeping, "disguised unemployment," and what kind of jobs have been created recently to keep the figures relatively low.

Professional economists, who once boasted about their discipline's ability to "fine-tune" the economy and their long-range forecasting accuracy, now are considerably more modest. Their track record as prophets in the past few decades has been less than impressive.

Economists once wrote textbooks in which they considered the theoretical problems to be faced: higher prices and inflation on the one hand, or higher unemployment rates on the other. They wrote and spoke of "tradeoffs" in these areas, and the politics of choice.

They were hard-pressed to explain "stagflation" when it first arose. How could the U.S. experience both double-digit inflation and double-digit unemployment? Previously this was considered impossible in their nicely developed theoretical models.

As the U.S. nears the dawn of the 21st Century, forecasts of improved conditions by government and academic economists are greeted with considerable skepticism by journalists, commentators and the general public. Others, on the far left wing of the political spectrum, argue that the nation's economic and social woes are due to the capitalist system.

Years ago Paul Sweezy, a leading Marxist economist, wrote in his book *Capitalism For Worse:*[1]

"The real question to which economists ought to address themselves...is why capitalism in the 20th Century has such a powerful tendency to stagnation that it requires increasingly massive forms of public and private waste to keep itself going at all."

The ability of an economic system to tolerate criticism and to adapt when necessary is vital to its survival. But the Joint Council on Economic Education has warned of "growing economic illiteracy" among Americans.

The U.S. has had a consistent tradition of listening to thoughtful criticism of the "system." Writer Studs Terkel, an analyst of blue-collar Americans, wrote:[2]

"Lately there has been a questioning of the work ethic, especially by the young...Unexpected precincts are being heard from in a show of discontent by blue collar and white ...On the evening bus the tense, pinched faces of young file clerks and elderly secretaries tell us more than we care to know... Ought there not to be another increment, earned though not received to one's daily work - an acknowledgment of a man's being?"

Attacks on the U.S. economic system sometimes have been made with great passion, but little reason. Although inflation has seriously eroded the value of the dollar and curtailed the U.S. standard of living, this problem is not uniquely American.

Nearly two decades ago Great Britain, for example, a historically prosperous West European nation, took comfort from 1977 statistics showing that her rate of inflation was down to 14%, a substantial improvement from the 30% rate of 1975.

Communist bloc nations, now defunct, once made much of their ability to command resources and avoid the so-called "anarchy" of capitalist market systems. Russia has experienced enormous problems of hyperinflation in its economic transition. After prices were decontrolled in January, 1992, prices on many items virtually quadrupled overnight. In 1992, inflation was 1353%. It was 896% in 1993, and still an estimated 292% in 1994, according to Russian authorities and the International Monetary Fund. On the eve of the Russian presidential election in the summer of 1996, older former Soviet citizens were talking about the "good old days" of Stalin.

The U.S. inflation rate, of paramount concern during the Carter presidency, now appears to be well under control. Yet economic problems are cyclical and inflation is a problem that constantly bears watching. In many other industrial nations, it is the most important economic challenge facing the government. In some cases in Latin America, for example, the cost of living has doubled in a single year.

The Problem of Scarcity

Many Americans, who live in the most productive and affluent society in history, may never even think of the central problem of modern economics: scarcity.

Scarcity relates to the relative shortcomings of our physical environment, the limitless desires of most human beings and the productive ability of the community.

Economist Robert L. Heilbroner says, "...If there were no scarcity, goods would cease to exist as a social preoccupation."

But all societies must deal with scarcity. They must struggle to survive, ultimately because of the scarcity of resources.

Wants and Needs

All human wants cannot be satisfied. We cannot have the proverbial cake and eat it too, however fervently we might wish to do so. Resources are limited, although our appetites are not. Some economists believe that a "rising tide of expectations," a distinctly post-World War II development, has led to many of the nation's problems.

Americans are constantly increasing their demands for goods and services. What was considered a luxury good during the 1950s often is considered necessary in 1996. Family budgets, as well as governmental ones, have grown bloated. Some people routinely overspend and several social critics have labeled the United States a "credit card civilization." Some regret the passing of the *"Puritan Ethic"* stressing hard work, thrift and savings.

Given scarcity, nation and people must make economic choices. How can the United States, for example, best use its scarce resources and its labor force most efficiently to produce those goods and services which will satisfy as many economic wants as possible? How can future production be increased? In a speech at Miami, Ohio, University, Walter B. Wriston, former chairman of Citicorp said:[3]

"There is no mystery about the definition of capital... Every economist from Adam Smith to Karl Marx agreed that capital is nothing but stored-up labor, either your own or someone else's...Somebody has to work hard enough to earn a wage and then exhibit enough self-denial to save some of what he earned...The grain bin holds the result of last year's labor. You can do two things with it; you can bake some bread and eat it; or you can use part of it to plant next year's crop. If you do the first, you have consumed your capital; if you do the second, you have invested it."

During the mid-1970s, President Jimmy Carter said that the nation faced a major energy "crisis," which he called the "moral equivalent of war." Americans, the Chief Executive said, had ravenous appetites for more and better creature comforts: cars, home appliances and many modern conveniences. The result at the time was the perception of a major scarcity of energy.

Presidents Ronald Reagan, George Bush and Bill Clinton have all faced economic issues of paramount importance on the public agenda.

How the U.S. copes with an energy "crisis" or other economic problems of scarcity is a test of the American tradition of enterprise, imagination and skill.

This slim volume will survey the evolution of capitalism, the philosophy behind it and some serious problems America currently faces. Alternative systems and their results also will be considered.

The central theme of the work is that *ours is a political economy*. Major economic decisions often are made for political reasons. What is good politics may be bad economics. Conversely, economists may be frustrated because politicians refuse to follow their advice. What is good economics, in that situation, may be bad politics.

In a nation of approximately 260 million people, it is not hard to find difference of opinion among both politicians and economists. And, of course, there are all kinds of economists and all kinds of politicians. Some see things from leftist perspectives, some are centrists and some are inclined to view policy choices from the right. Debate about politics and economic policy making is often fierce and heated. Sometimes it is even enlightening.

The public official's perspective also depends upon his or her office. Presidents have, or should have, a national perspective: What is in the best interest of the U.S. and its people? Senators tend to think of their states, or in some cases, their regions. Members of the House often think first and foremost of interests in their local districts.

When President Bill Clinton, a Democrat, urged Congress to approve the North American Free Trade Agreement (NAFTA), many members of his own party, including House leader Richard Gephardt, refused to support him. Not a single Democratic representative or senator from Michigan voted with the President. Had it not been for Republican congressional support, NAFTA would have died.

Democratic senators and representatives from constituencies with strong labor unions, such as the I-75 corridor in Michigan, were thinking first and foremost of their political base. Without labor support, they may not be able to get re-elected. They would not put it that way, of course.

In this media age, politicians and interest groups (including unions) hire expensive public relations advisers to put the correct "spin" on their motives. Votes are always motivated by the pure national interest - as they see it. The unstated view of many, if not most, members of Congress is that the national interest requires that they be re-elected!

Political scientists have made distinctions between trustee and delegate perspectives of representation. The "trustee" view, voiced long ago by British Conservative Edmund Burke, holds that a member of Parliament owes nothing to his constituents but his own best judgment. The delegate view, called by some cynics the "errand boy" (or girl) perspective, holds that one should vote according to the views of one's own constituents. Forget that and you will be a FORMER member of Congress.

***

II. COMPARATIVE SYSTEMS: TRADITION, MARKET AND COMMAND

There have been many systems of exchange throughout our economic history, ranging from the barter system to our own age of money. And money, of course, is anything society is willing to accept as a medium of exchange. For purposes of convenience, some contemporary scholars are inclined to divide systems into three categories: tradition, market and command.

Although professional economists tend to be more interested in the market than in any other kind of organized economy, most human beings live in "traditional societies." Such peoples and nations are primarily of interest to the professional anthropologist.

Because command economies are virtual extensions of the state, they are of greater interest to the political scientist than to the economist. The latter, however, cannot ignore command systems, given their considerable power and influence in the 20th Century.

A. Traditional Economies

In an economic sense, traditional societies are characterized by the following:

(1) *An overwhelmingly agrarian labor force.* As many as 70% to 90% of the workers typically earn their living on the land.

(2) *A static system of production and distribution.* With large "hidden surpluses" of peasant laborers, many of these traditional systems produce remarkably little.

Although some nations, such as India, claim that the so-called "Green Revolution" with its huge irrigation projects, new and better chemical fertilizers and agricultural modernization place them on the verge of self-sufficiency, others are skeptical about such claims. India's rapidly-growing population, some believe, literally may eat up all the increased food production.

"If our population continues to increase as rapidly as it is doing," said Mohammed Ayub Khan, President of Pakistan several revolutions ago, "we will soon have nothing to eat and will all be cannibals."

(3) *Inefficient units for agricultural production.* Developing nations suffer from what some economists call "postage stamp cultivation." They have small units of agricultural production that are too small and too inefficient to use modern methods and equipment efficiently.

(4) *Overwhelmingly poverty.* Although some traditional societies, such as Kuwait or Iran, are not poor, most of them are. In 1960, the average Gross Domestic Product (GDP) in the developed world - U.S., Canada, Europe, Oceania, Israel and Japan - was $6520 (U.S. dollars). In the Third World, the average GNP was $361, or $6159 less than the rich countries.[4]

By 1988, rich nations enjoyed an average per capita income of $13,995 and the poor countries $717, resulting in a gap of $13,278, an increase of $7119 over the 28 year period. Thirty years ago, economist Barbara Ward, in her book, *The Rich Nations and the Poor Nations,* set a dividing line between the two. Nations with less than annual per capita incomes of less than $500 were poor. In this category are India, Pakistan and Egypt, among others.

Distribution of World Population and Gross National Product Per Capita in 1991 showed that citizens of Switzerland had the highest per capita income in the world at $33,610, followed by Japan ($26,930), Germany ($23,650) and the U.S. ($22,240). At the other end of the list were nations such as Zimbabwe ($650), China ($370), India ($330), Burundi ($210) and Mozambique ($80).[5]

Third World poverty may be difficult for Americans to comprehend. It is not the kind of poverty with which they are familiar, that of the urban ghetto. It is much worse, with starvation common. Television reports from Somalia or Sudan show the horrible reality of life in these areas.

In all, about one billion people live in want, lacking access to basic nutrition or health care. Two-thirds of them live in the densely populated states of India, Bangladesh, Pakistan and Indonesia.

(5) *Rural isolation.* Most traditional societies are isolated and rural in character. The national government is often seen as something that is remote and distant. It has little to do with the peasant's daily lives, as they see it. Clearly, it does not command their paramount loyalty which is to the family or clan.

Political scientists and anthropologists report that peasants believe the less one has to do with government, the better. Rural villagers often feel that their primary roles in life are to provide sons for the military and taxes for the government.

(6) *Undemocratic governments.* Given these social and economic conditions, one might expect, quite correctly, that democracy would be rare in such nations. One-party rule is common.

"Economic instability is the reason so many of our nations have dictatorships," said former Venezuelan President Carlos Andres Perez in an interview with *U.S. News & World Report:*

> "In general, our misery, our lack of education, our malnutrition diminish our creative capacity. This stirs up conflicts and tensions which, in turn, lead to dictatorships being created in order to restore order. The only solution to this tragic cycle is to establish just and equitable international economic relations, and this directly involves the industrialized nations-and especially the United States..."

Illiterate peasants struggling to subsist are usually too preoccupied to engage in politics.

(7) *Inadequate health care delivery* also is a major economic and social problem in traditional societies, as statistics from such organizations as the World Health Organization (WHO) indicate. Infant mortality is high, life expectancy is low. Data on life expectancy is particularly important. In 1950, citizens of low-income countries lived an average of only about 35 years. The average life expectancy in rich countries was about 65. By the early 1970s, the life span in Third World nations was about 55 while it had reached 71 in rich countries. In 1991, the difference between rich nations and poor nations was about 12 years. The United Nations Demographic Yearbook shows, for example, life expectancies of 44 years in Angola, 48 in Nigeria, 47 in Tanzania and 55 in India.[6]

Third World citizens suffer and die from malnutrition. Inadequate supply of healthy and nourishing food kills more people than does starvation, although it may make less dramatic pictures on the network evening television news. Still another serious problem is the lack of safe and clean drinking water. About half of the people in the world face this cruel reality.

A child dies as a result of malnutrition every three seconds somewhere in the world. This adds up to 1000 per hour, 30,000 a day and a total of 10 million a year.[7]

What should America's relationship be to these nations? Should the U.S., as many suggest, extend further economic and technical aid to help these people with the task of building a modern society? If so, what, if any conditions, ought to be attached to American aid? Should the United States withhold aid from those nations which choose to follow command models of economics?

Or should the U.S. give priority to its own domestic economic problems, particularly given the often strident and anti-American tone of statements made by "Third World" nations in the United Nations General Assembly?

> "The modern age and our common morality insistently demand respect for human dignity...But a world in which poverty and misery continue to afflict countless millions would mock these imperatives. The daily preoccupation of men and women would be the harsh necessities of survival; the energies of nations would be consumed in hatred and rivalry. The United States proceeds from the conviction that both morality and practical interest point in the same direction, toward a dedicated enterprise, of cooperation...So let us get down to business... We owe our people performance, not slogans; results, not rhetoric."
>
> --Henry Kissinger, former U.S. Secretary of State

In which direction will the "traditional" societies go? Will they emulate American political and economic democracy? Or will they adopt a command-style system similar to that of the former Soviet Union?

In the realm of economics, many Third World nations "demand" redistribution of the world's riches and advocate cartels to control supply and prices of such commodities as sugar, copper, coffee, bauxite and, of course, oil.

Some scholars see this conflict as essentially a matter of the "have" nations of North America, the United States and Canada, Western Europe and Japan versus the nations of Latin (South and Central) America, Asia, Africa and the Middle East.

To most Third World governments, the lesson of the 1973 embargo by the Arab members of the Organization of Petroleum Exporting Countries (OPEC) on shipment of their oil to the United States, the Netherlands and Denmark is clear. The West, they insist, must pay more for their resources in the future.

Whether such policies will enable these nations to modernize and develop rapidly remains questionable. Some students of economic development believe that Western foreign aid, trade and investment may be curtailed as a result of these moves.

The fate of the United States and the world, however, will be significantly influenced by the direction taken by these nations. They are trying, with varying degrees of success, to break the stranglehold of tradition on their economic and social development.

> "Development, despite all the efforts of the past 40 years, has failed to close the gap in per capita incomes between the developed and developing countries - a gap that at its extremes ranges in money terms to more than $8000 per capita...The proposition is true. But the conclusion to be drawn from it is not that development efforts have failed, but rather that closing the gap was never a realistic objective in the first place...What is far more important...is to seek to narrow the gaps...in terms of the quality of life - in nutrition, literacy, life expectancy, and the physical and social environment...Is that feasible? It is."

--Robert S. McNamara, former president of the World Bank

B. The Command Economy

Contemporary economists would generally agree that there no longer is such a thing as a "pure economy." Most nations in the modern world have "mixed" systems. But it remains true that most industrial economies are organized primarily around either the principles of free enterprise or the principles of government command.

Free enterprise, or capitalism, relies on private citizens to produce the goods and services needed by society. It gives them the right to own the means of production and relies heavily on their initiative as well as the impersonal forces of the market to regulate the economy. The market allocates resources and establishes "appropriate" levels of rent, prices, wages and profits.

"...Economic freedom is also an indispensable means toward the achievement of political freedom," asserts economist Milton Friedman in *Capitalism and Freedom*. "The kind of economic organization that provides economic freedom directly, namely competitive capitalism, also promotes political freedom because it separates economic power from political power."[8]

The principal alternative to capitalism is government command or socialism. Theoretically, command economies give government the power to plan and control nearly all productive activities. In such nations, the former Soviet Union being a good example, the economic system tends to be an extension of the state.

But although such systems have denied the right of private ownership of the means of production and have stressed public planning of production rather than the market system, they seldom are identical.

The term, "*command economy*," describes those systems using totalitarian political methods, as well as state-directed economies.

England, under a Labour Party government, often was referred to, quite incorrectly, as a "socialist" state. Although the Labourites were committed to aspects of planning, they had little interest in nationalizing much more of British industry, presently about 20%-25% of the nation's economy.[9]

In the 1960s and early 1970s, Britain appeared to have reached a consensus on many issues of political economy. That brief tranquility was ended with the advent of Prime Minister Margaret Thatcher and her anti-welfare state policies.

Income and estate taxes were cut. A number of nationalized state industries were privatized, including steel, gas, airways and telecommunications. The power of trade unions was curbed. But Thatcher was unable to reduce the size of the welfare state itself significantly.

The American media have given very little news space or air time to cover British economic policy differences. This may be understandable, given the domination of British politics by the Conservative Party under Prime Ministers Margaret Thatcher, elected three times (in 1979, 1983 and 1987) and John Major, elected in a close 1992 contest. That changed with the election of Tony Blair, a Labourite, as prime minister on May 1, 1997.

Britain and other Western European nations, often labeled "Democratic Socialists," are committed to some government direction of the economy, but always with public consent through the democratic process. In short, they are democrats first and socialists second.

Social Democrats reject Marxist notions of the inevitability of violence and revolution as the vehicle of social change. In most instances, they have produced a "hybrid" economy - part socialist, part capitalist.

"...If Labour cannot obtain a majority, it must as a minority accept the will of the majority," wrote Clement Atlee, first Labour Prime Minister of Britain after World War II.

Other "hybrid" economies unlikely to fit theoretical "models" of either "pure" free enterprise or "pure" socialism would be those of Japan, most of the less developed countries (LDCs) and even France.

The roots of socialism are buried deep in history. But many economic historians believe that it developed as a major force in reaction to modern capitalism and the demands of the Industrial Revolution.

Karl Marx is the most influential of the fathers of socialism. Marxists insist that since Communism comes only with the "withering away of the state," no nation has yet achieved the Communist Utopia. However, the term Communism is used loosely by most people to refer to those nations such as the former Soviet Union and the People's Republic of China, which have had such visionary blueprints.

Marxists protested the injustices of 19th Century European capitalism. They rejected wealth as a measure of worth and the use of money as a yardstick to measure human values. Marx denounced "exploitation " of labor, social injustice and profit. He gave expression to the theory that labor alone was the source of value, an idea accepted today by few non-Marxist economists.

Marx believed that capitalism was its own gravedigger, since it created an urban working class (proletariat) which would rise up to destroy it, once its lot became intolerable and it realized its revolutionary power.

"The development of modern industry...cuts from under its feet the very foundation on which the bourgeois produces..." Marx and Friedrich Engels wrote in The *Communist Manifesto* in 1848. "Its fall and the victory of the proletariat are equally inevitable."[10]

The Revolution would destroy capitalism and end class struggle. Human nature would eventually change. Man would become cooperative, rather than competitive. After a vaguely defined period of a "dictatorship of the proletariat," the state ultimately would disappear because men would become angels.

But Marx's crystal ball was, in many respects, cloudy. Many of the objectives of the Communist Manifesto and his other revolutionary writings are today accepted benefits of capitalist, democratic states.

Marx failed to visualize the important role that labor unions would play. If he could be "resurrected" from his grave in 1996, he would hardly recognize the capitalism he damned.

Today workers under capitalism enjoy paid vacations, minimum wage laws, shorter work weeks, profit sharing, pensions and many other "fringe" benefits.

Marx also was mistaken in his conviction that the revolution would occur first in the most advanced industrial nations, since "capitalism must run its course," as other historical systems had.

When the Bolsheviks seized power in Russia in 1917, they attempted, with chaotic results, to implement some of Marx's ideas. Lenin stopped this in 1921 by introducing a *New Economic Policy (NEP)*. It was intended to be a partial retreat to capitalism in order to buy time, avoid starvation and stimulate production.

It was Josef Stalin who, in 1928, began the *"Socialist Offensive."* Stalin used government command to force rapid industrialization and collectivization of peasant agriculture. Millions who resisted were liquidated.

Under Stalin and his successors, the former Soviet Union used its command system of economic organization to build itself into a major world power. But five- and seven-year national economic plans modernized Russia at a heavy social cost, creating a police state, restrictions on human rights and, until the latter years of its existence under Mikhail Gorbachev, the worst sort of terrorism.

Under a command economy, the Soviet government as well as those of its Eastern European allies, determined what was produced and in what quantity, what methods of production would be used and what prices would be charged. After 1989, called by the media "The Year of Falling Dominoes," Eastern Europeans abandoned their command systems to adopt free market economies. The new Russia soon followed suit.

As noted, this transformation has been painful. Problems of unemployment, inflation and uncertainty are common. In Russia the future of democratic reforms, as well of the market system, is not at all assured. Some news media in the summer of 1996 described Russia as a nation threatened with complete social disintegration.

Why would Russians or Eastern Europeans be willing to pay the painful social costs of transition from command to market economies? Perhaps it is because free market systems promise benefits which command economies never have been able to deliver in any sustained way. But will the Russians have patience in their time of trouble, or revert to some form of totalitarian socialism?

Looking back at more than 220 years of freedom, many Americans conclude that our present economic system has contributed greatly to the development of liberty. They reject "command" economics for our industrial society and view contemporary proposals for a greater economic role for government with suspicion.

They may be inclined to see "command" economies as "coercive" economies. Economic freedom is an important part of liberty and, in their judgment, that may well be best promoted by free enterprise.

C. The Market Economy

Capitalism, or free enterprise, is a system of economic organization in which individuals own land, housing, farms, factories and shops - small and large.

In economists' terms, *individuals,* singly or collectively - in partnerships or through corporations - *own the means of production*. They may pursue profits in using their labor, money (capital) or natural resources.

Free enterprise traditionally has relied heavily on the *impersonal mechanism of the market* to determine the most *efficient allocation of resources*. The market is a structured environment in which buyers and sellers bid freely among themselves to determine wages, rents and profits. In all markets, it is this *interaction of buyers and sellers that determines what is produced, how it is produced, who receives the output of production and what prices they pay.*

Adam Smith, explaining what a market is, wrote in *The Wealth of Nations*:[11]

"When the quantity of any commodity which is brought to market, falls short of the ...demand, all those who are willing to pay the whole value of the rent, wages and profits...cannot be supplied with the quantity which they want. Rather than want it altogether, some of them will be willing to give more. A competition will immediately begin among them and the market price will rise more or less above the natural price. ...The interest of all other laborers and dealers will soon prompt them to employ more labor and stock in preparing and bringing it to market. The quantity brought thither will soon be sufficient to supply the effectual demand."

The social effect of all this, free enterprise theorists contend, would be good. Resources would be allocated efficiently and productively. A high degree of stability would occur at efficient levels.

The consumer was *"sovereign,"* casting dollar votes. Perhaps best of all, this efficient system would be achieved, Smith believed, without coercion. Government need not be involved.

Smith and others advised that government should follow a policy of *laissez-faire* (hands off) with respect to economic activities of private individuals.

"It is not from the benevolence of the butcher, the brewer or the baker that we expect our dinner," Smith wrote, "but from their regard to their self-interest. We address ourselves not to their humanity, but to their self-love, and we never talk to them of our necessities, but of their advantages."

Smith and two prominent 19th Century economists, David Ricardo and Thomas Malthus, believed that a nation's economy depended on the *three factors of production*: land, labor and capital. These "factors" *would be kept in balance by the marketplace, the chief regulator of supply and demand.*

Modern capitalism still makes these assumptions, although problems of the industrial age, notably the rise of oligopoly (domination of an industry by a handful of large firms) causes some to question the validity of these classic promises on the eve of the 21st Century.

"People of the same trade seldom get together," Smith wrote in a passage business leaders tend to forget, "but the conversation ends in a conspiracy against the public, or in some diversion to raise prices."

The United States has become an economic colossus under free enterprise, enjoying a standard of living that is the envy of most of the world. It has rejected socialist solutions. Ironically, however, many of the social problems that accompanied earlier capitalism and were denounced by Marx so passionately, have been corrected by legislation.

But free enterprise has been compatible with American political goals, particularly the stress on individualism and freedom.

Smith's thinking has had a profound effect on American economic life and although much of what he said may be viewed as oversimplified today, his influence lingers on, for many believe that the core of what he said remains true.[12]

Some contemporary political economists debate the relationship of economic systems and political freedom. A leading conservative, Professor Milton Friedman, has written: "The existence of a market mechanism separate from the state provides the necessary basis for a free political life. Existence of capitalism is a necessary condition for political freedom." But Friedman, perhaps the leading exponent of *laissez-faire* today, is at odds with those who believe that the real strength of American capitalism lies in its adaptability.

Clearly, Adam Smith would not recognize contemporary American capitalism. He assumed *equilibrium,* but we experience sharp cyclical variations, including inflation and/or depression. He assumed pure competition, but we witness a decline in competition with the development of industries in which a few firms dominate *(oligopoly).*

John Maynard Keynes' idea that government deliberately should incur budgetary deficits to stimulate recovery, is sheer "heresy" to free-enterprise purists. Keynes, however, did devise his theories in response to the Great Depression and for the purpose of rescuing the capitalist system.

Even some conservative historians have referred to Franklin D. Roosevelt as the "savior of American capitalism." Considering the American system before and after the Great Depression, Roosevelt may well have been the father of *"Welfare Capitalism,"* a very different, but perhaps more adaptable system for 20th Century America.

Marxists commonly believed that the Great Depression was the prelude to the collapse of capitalism. But the disasters they predicted failed to materialize. Although moderate recessions have occurred, they generally have not lasted very long.

Consequently, the new "Welfare Capitalism," regulated as Adam Smith never would have predicted, has proved to be an efficient way to stimulate the productive power of the American people. Liberals and Conservatives, Democrats and Republicans, as well as those of other ideological bents, all debate the extent to which government ought to intervene in the economy. But few of them have any desire to kill the proverbial "goose that laid the golden egg."

***

III. CAPITALISM AND HUMAN BEHAVIOR

If one were to poll a group of adult Americans as to the meaning of capitalism, one probably would get some strange answers. If one were to ask what assumptions free enterprise makes about human nature, one might get even wilder replies.

Yet it is important to understand the American system of political economy. Some 20 years ago American Democratic Capitalism marked its bicentennial. The year 1976 marked not only the 200th anniversary of the publication of Adam Smith's *Wealth of Nations,* but also the Declaration of Independence.

Capitalism began to develop in 18th Century Britain as a part of the movement of individualism. Later it spread throughout northwestern Europe and the United States. Its basic principles were clearly stated by British philosophers, Smith, David Ricardo, Thomas Malthus and John Stuart Mill.

These free enterprise philosophers assumed that man's needs were primarily individual, rather than communal. Individual freedom could best be promoted, they reasoned, by granting ownership of the means of production (land, labor and capital) to private citizens rather than to the state. This economic system would be most harmonious with democracy because *it would diffuse economic power, just as democracy has diffused political power*.

"There is an intimate connection between economics and politics," asserts Milton Friedman, one of today's free-enterprise philosophers and Nobel laureate in economics, in *Capitalism and Freedom:*[13]

> "Economic arrangements play a dual role in the promotion of a free society. On the one hand, freedom in economic arrangements is itself a component of freedom broadly understood, so economic freedom is an end in itself. In the second place, economic freedom is also an indispensable means toward the achievement of political freedom...The kind of economic organization that provides economic freedom directly, namely, competitive capitalism, also promotes political freedom because it separates economic power from political power and in this way, enables the one to offset the other."

Just as government could be oppressive in a political sense, so a state-directed economy could control other vital aspects of people's lives. Economic freedom, capitalist theorists believed, would also ensure economic and technological progress. Why?

Man, they believed, is naturally acquisitive and competitive. He works best when he has incentive and when he works for himself. Capitalism is a powerful engine, fueling the economy because it is *both a profit system and a loss system.* Individuals thus are strongly motivated to put forth their best efforts to succeed, lest they be forced to pay the price of failure.

The late Bernard Baruch, advisor to several Presidents, once expressed this view of human nature in a college commencement address when he said:

"The moving forces of mankind are acquisitiveness, the urge to function as an individual, a yearning for freedom in mind and body, and above all the constant quest of opportunity to advance. These are the attitudes of individualism and the man without them is not worth his salt. We can't repeal human nature by an act of Congress."

American society quite clearly values competition, whether on the playing field or in the marketplace. The desire to excel, to be "number one" is manifested daily in our lives.

But among the philosophy of free enterprise economics, there is no unanimity on the role of government. Smith himself assigned to the government responsibility for education as well as for national defense.

Wasily Leontieff, Nobel Prize laureate in economics and former professor at Harvard University, once wrote:[14]

"The pursuit of economic gain is certainly the mighty power source that propels the American economy forward...The profit motive in particular promotes and safeguards its unequaled technical and managerial efficiency...But to keep it on a chosen course, we have to use a rudder. The steering apparatus consists of taxes, subsidies, anti-pollution regulation, and other measures of governmental economic policies. There are...those who say that we should simply hoist the sails and let the vessel go before the wind in whatever direction the wind happens to be blowing. This type of navigation is bound to bring the ship off course and land it on the rocks."

The molders of free enterprise also have lauded freedom of consumer choice. Citizens are free, they noted, not only to cast their "dollar votes" for those goods and services they wish to acquire, but also ultimately to determine what will and will not be produced.

"Consumption is the sole end and purpose of all production," wrote Smith in *The Wealth of Nations.* "The interest of the producer ought to be attended to only so far as it may be necessary for promoting that of the consumer."

Vital to capitalism is competition and the early capitalist thinkers exalted it. It was assumed that many small producers would vie fiercely for consumers' favor. This ultimately would result in the most efficient use of resources.

But history has shown that pure competition declined with the Industrial Revolution and the subsequent rise of both giant industry and giant labor unions. Many markets are dominated by a few firms, a condition called oligopoly, and there is a widespread suspicion among some Americans that price collusion and anti-competitive practices prevail in at least a few of these industries.

"It is now considered the sense of the community, even of economists...," said John Kenneth Galbraith in an address as president of the American Economics Association, "that the most prominent areas of market oligopoly - automobiles, rubber, chemicals, plastics, alcohol, tobacco, detergents, cosmetics, computers, bogus health remedies, space adventures - are areas, not of low but of high development, not of inadequate but of excessive resource use."

Galbraith, an iconoclastic economist, argued that the economic might of major corporations gives them a political and social power not considered in classic economic theory. He contended that social needs such as "housing, health services and local transportation" suffer from deprivation.

Says Galbraith, "The defender of established (economic) doctrine does, of course, argue that excess and deprivation in resource use in the areas just mentioned reflect consumer choice...The explanations beg two remarkably obvious questions: Why does the modern consumer increasingly tend to insanity, increasingly insist on self-abuse? And why do little monopolies perform badly and the big ones too well?"

Although the American tradition of free enterprise views government intervention in the economy with suspicion, intervention to foster competition, to meet social needs and to deal with unemployment and emergencies such as war, have become facts of our history.

Federal use of fiscal policies recommended by British economist John Maynard Keynes at the time of the Great Depression have become important tools of economic stabilization. Interest rates, taxes and Federal spending levels have become permanent topics of American political and economic debate.

IV. THE ROOTS OF AMERICAN CAPITALISM

The roots of modern American capitalism sprouted in Great Britain, as did the roots of U.S. political institutions. We inherited from England not only the political values of representative government, the limited state, individualism and civil liberties, but also the new technology of the Industrial Revolution and the ideology which nurtured its development.

The British belief that contracts and property rights were "sacred" found the New World environment quite receptive. One of the most important of all American beliefs brought from England was the "Protestant Ethic," a set of ideas and values which have affected Western Civilization since the early days of capitalism.

The "Protestant Ethic" stressed such values as *individual responsibility, thrift and savings*. Spending all that one had earned was regarded by church leaders and their congregations as immoral. Success was considered an indication of moral worth and virtue, just as failure was interpreted as a sign of poor character. Hard work was "good for the soul" and retirement, unless caused by ill-health, was viewed as a sign of laziness.

Russell Conwell, noted Baptist minister and president of Temple University in Philadelphia, was the author of one of the most popular sermons of the late 19th and early 20th Century. Called "Acres of Diamonds," it was given some 6000 times over a period of 50 years. The Rev. Conwell said:[15]

"It is an awful mistake...to think you must be awfully poor in order to be pious... Some men say, 'Don't you sympathize with the poor people?' Of course I do...But the number of poor who are to be sympathized with is very small...Let us remember that there is not a poor person in the United States not made poor by his own shortcomings... It is all wrong to be poor, anyhow."

Conwell's view was echoed by the Episcopal bishop of Massachusetts, one William Lawrence, who in 1890 commented:[16]

"In the long run, it is only to the man of morality that wealth comes...Godliness is in league with riches...Material prosperity is helping to make the national character sweeter, more joyous, more unselfish, more Christlike."

The dominant early American political and economic ideologies were compatible. Thomas Jefferson's Declaration of Independence and Adam Smith's *The Wealth of Nations,* both published in 1776, stressed the paramount importance of the individual and the need for minimal government.

The early U.S. economy was a relatively simple one. Most Americans were farmers and it was said that democracy, to be successful, required a nation of free, small independent farmers, working land that they could call their own.

Jefferson's vision of America's future emphasized the moral superiority of rural life. Cities, he feared, would jeopardize the health of the body politic. He wrote: "The mobs of great cities add just so much to the support of pure government, as sores do to the strength of the human body."

Gradually, however, the agricultural economy shifted to an industrial one. Jefferson's self-reliant small farmer began to raise crops, not only for his family, but also for the market.

In time, the imperatives of a new agricultural technology and the requirements of economic efficiency gave rise to the large corporate farms. The family-sized farm, unable to take advantage of economies of scale, or to afford the new technology, often came in for hard times.

With the rise of a factory system came *urbanization.* Farmers flocked to the cities where they competed with immigrants for jobs, housing and other necessities. This heavy rural-urban migration, coupled with early unrestricted immigration, swelled the population of U.S. cities and created subsequent social problems previously unknown in the New World.

Today more than 90% of the American people live in or near metropolitan areas. The manufacturing process, which long ago created an efficient division of labor with workers performing only a small role in producing a finished good, has grown increasingly complex. We live in an age of *automation,* where use of machinery is most efficient in large-scale operations.

Computers and *industrial robots* may have been beyond the dreams of even the highly imaginative and intellectually-gifted Jefferson.

Zbigniew Brezinski, former Columbia University professor and national security adviser to President Jimmy Carter, wrote in *Between Two Ages*:[17]

"Man has always been involved in a grubby, daily struggle against his environment...Until now, man has lived in combat with nature. Man has won that war...We are now, more or less, standing on the battlefield...wondering what comes next."

Despite industrialization and urbanization, despite the move toward what Brezinski calls the *"technetronic society,"* characterized by electronics and computers, many Americans have maintained their faith in traditional values.

Others seriously question the "relevance" of rural values in a modern, complex, industrialized nation.

What values in particular are under attack from critics of traditional American beliefs? Perhaps foremost is self-reliance and individualism.

Some have said that self-reliance and individualism were nurtured primarily by the American frontier which, according to the U.S. Bureau of the Census, passed into history with its closing in 1890.

Historian Frederick Jackson Turner believed that the frontier was the *"safety valve"* in U.S. economic development. Those who failed in the East always could pack up and start anew as pioneering farmers. The kind of virtues required to survive the harsh frontier environment, independence and self-reliance, may no longer be needed, according to some critics of the economic system.[18]

Capitalism should be reformed, they argue, to stress cooperation, sharing and redistribution of market-determined incomes. These critics contend that contemporary capitalism produces many social casualties, who for a variety of reasons, simply cannot compete successfully. Others suggest that this had more to do with modern industrial societies than with a particular kind of economic system.

Ideology, vital as it is to the process of development, is only a partial explanation to the success of the enterprising Americans.

We have been blessed with an abundance of natural resources, fertile farm land, temperate climate, vast forests and mineral wealth. We have produced an educated labor force whose record of skill and productivity is remarkable. Private enterprise, encouraged by government, has developed these resources to make the U.S. the world's leading industrial power.

But many of the undesirable effects of the Industrial Revolution remain with us. Deeply-rooted conflict between rural and urban areas, between agricultural, industrial and union interests, have become imbedded in American life.

***

V. CONSEQUENCES OF THE AMERICAN INDUSTRIAL REVOLUTION

Before the Civil War, the U.S. was primarily a rural agricultural nation, a society in which individual opportunities were great.

In case of failure, a farmer always could pack up his belongings and move west. But in 1890, the U.S. Bureau of the Census announced the closing of the frontier. This important "safety-valve" opportunity for the individual was no longer available.

With opportunity in the West thus restricted, the farmer who failed to make an adequate living in rural America had little choice. He usually moved into the emerging city where the dream of Alexander Hamilton - an urban, industrial powerful nation - was becoming a reality.

As George Washington's Treasury Secretary, Hamilton had proposed that government aid business by giving bounties to new enterprises. In 1791, Hamilton argued forcefully that manufacturing was vital to U.S. security and prosperity.[19]

Said Hamilton, "...Not only the wealth but the independence and security of the country appear to be materially connected with the prosperity of manufacturers... Every nation with a view of those great objects, ought to endeavor to possess within itself, all the essentials of national supply."

After the War of 1812, Thomas Jefferson accepted the view that manufacturing growth and development were essential.

"We have experienced what we did not then believe...that to be independent for the comforts of life we must fabricate them ourselves. We must now place the manufacturer by the side of the agriculturist," he declared.

The 10-minute Bessemer process of making steel from pig iron was a significant technological development of the period. The discovery, generally attributed to England's Sir Henry Bessemer, used air in place of charcoal as a fuel in the chilling of iron. Oxygen united with the impurities of iron leaves behind the pure iron.[20]

Andrew Carnegie of Pittsburgh was one of the first of America's industrial leaders to use the Bessemer process. Workers went on a then-unusual system of three eight-hour shifts. Carnegie became convinced that workers could not labor for 12 hours in the "little hells" of steel manufacture with efficient results.

Carnegie celebrated the growing industrial power of the U.S. and his own growing fortune, in a classic essay entitled *Wealth,* published in *North American Review.*[21]

"Today the world contains commodities of excellent quality at prices which even the generation preceding this would have deemed incredible," he marveled. "The poor enjoy what the rich could not before afford. What were then luxuries have become necessaries of life."

John D. Rockefeller and his Standard Oil Company became even richer with the development of new oil drilling methods. From its modest beginnings in the 1860s, Standard Oil developed by 1900 into a giant which owned more than 80% of U.S. crude oil refining capacity. Rockefeller later reflected on his hard work which gave his firm such a pre-eminent position. He wrote:

"...How often I had not an unbroken night's sleep, worrying about how it's all coming out. All the fortune I have made has not served to compensate for the anxiety of the period. Work by day and worry by night, week in and week out, month after month."

Eli Whitney's prewar discovery of the principle of interchangeable parts was beginning to put more furniture and appliances within reach of more American families.

By 1914, Henry Ford, who had created a sensation in 1908 with his Model T auto, was producing cars on an innovative assembly line.

From fourth place among the world's leading industrial nations prior to the Civil War, the U.S. rose to first by 1894. By 1914, she was producing as much as her three nearest competitors - England, Germany and France - combined.

Mechanization was everywhere. Slow, costly, inefficient efforts of craftsmen were unable to compete with rapid, less expensive, more efficient machinery.

To those who complained, Carnegie answered:[22]

"The price we pay for this salutary change (increased productivity and higher living standards) is, no doubt, great. We assemble thousands of operatives in the factory...of whom the employer can know little or nothing, and to whom the employer is little better than a myth. The price which society pays for the law of competition...is also great, but the advantages of this law are also greater still, for it is to this law that we owe our wonderful material development."

By 1875, even shoemaking was highly mechanized. Most Americans today probably do not know that shoes designed specifically for the left or right foot were not the rule in early America.

Meat packing and canning advances made possible shipments of perishable fresh beef in iced railway boxcars. By 1890, housewives could buy over the counter virtually everything that their mothers used to "put up."

Another important technological change was in metal-working machinery. Within 25 years after the South's surrender, Cincinnati was bidding for first place in machine tool manufacturing. Before World War I, it was on its way to becoming the most important center of machine tool production in the U.S.

The impact of the U.S. Industrial Revolution was felt on the farm as well as in the factory. Traditional hand tools were replaced and labor saving devices such as the McCormick reaper and John Deere's steel plow were introduced.

But increases in food production brought on by cultivation of new, fertile land and improved technology created an oversupply problem.

Economists tell us that *demand for food is "inelastic,"* that is, we can increase our consumption only so much. Unlike the manufacturer, who can respond to reduced demand by cutting production, the farmer aggravated his problem by producing still more in an effort to prop up his falling income.

The era from the Civil War to 1894 often is called "The Dark Age of American Agriculture." Parties of rural protest - Grangers, Populists, Greenbackers - voiced their demands for a better deal for the farmer. Hard times persisted until 1896, when the farmer entered a period - until 1915 - called the "Golden Era" of American agriculture.

***

VI. GOVERNMENT AND BUSINESS IN THE GILDED AGE

From the end of the Civil War to World War I, America built its tremendous industrial might and opened heated debate over government policy toward business.

The era often has been called the Gilded Age, a phrase unwittingly given to it by Mark Twain and Charles D. Warner who wrote a book of that title in 1873.

Government was primarily a promoter and friend of industry, rather than a regulator. Business taxes were deliberately kept low to stimulate investment and economic growth. The principal sources of Federal revenue before the 16th Amendment created income taxes were tariff and excise taxes.

Congress imposed tariffs to protect *infant U.S. industries* and to *promote economic expansion* as early as 1816. Partisan conflict, pitting *free-trade* Southerners against Northern *protectionists,* persisted until the Civil War. Protectionists boosted rates significantly in the postwar era, stirring up debate again.

Congress subsidized construction of the railroads, doling out land grants and loans to stimulate a modern, national transportation system.

Many sectors of American society supported industrial capitalism during the Gilded Age. Protestant clergymen Russell Conwell, Henry Ward Beecher and William Lawrence told their congregations, "Godliness is in league with riches."

Social philosophers, many of them infected by the gospel of Social Darwinism, added their support to the captains of industry.

Yale sociologist William Graham Sumner strongly believed in the survival of the fittest.[23]

He argued, "...if we do not like the survival of the fittest, we have only one possible alternative, and that is the survival of the unfittest."

Supported by churchmen, scholars, political and government leaders, American businessmen became the most admired types of the era. They were quick to defend themselves and the system against assaults of hostile politicians and would-be reformers.

Carnegie, who started out as a poor Scotch immigrant and became one of the wealthiest men of his age, wrote: "The price which society pays for the law of competition is great...it is here; we cannot evade it. No substitutes for it have been found."[24]

John D. Rockefeller, whose name is synonymous with great wealth, said the best philanthropy "...is not what is usually called charity. It is money carefully considered with relation to the power of employing people at a remunerative wage, to expand and develop the resources and to give opportunity for progress and healthful labor where it did not exist before. No mere money-giving is comparable to this in its lasting and beneficial results."

Not all Americans were willing to accept as "natural" or as "inevitable" the *growing concentration of economic power.* While America greatly admired the enormous efficiency of its productive machine, it was sometimes deeply skeptical about its industrial and financial leaders.

Historian Henry Adams, who observed that "...the world, after 1865, became a banker's world," was highly critical of Carnegie's theory that "natural selection" placed some in positions of tremendous wealth.

Adams countered: "The progress of evolution from President Washington to President Grant was alone evidence enough to upset Darwin."

Many middle-class Americans - small businessmen, professionals, white collar workers and others - feared the growing abuse of economic power from above and, even worse, the surge of "radical" movements, such as populism and socialism, from below.

Theodore Roosevelt reflected this middle class concern in a 1906 letter to William Howard Taft. "The dull, purblind folly of the very rich men; their greed and arrogance...corruption in business and politics, have tended to produce a very unhealthy condition of excitement and irritation in the popular mind, which show itself in the great increase in socialist propaganda," Roosevelt wrote.

Roosevelt believed the proper role of government was not to destroy trusts or businessmen. He wanted, he said, only to limit power of those who competed unfairly:[25]

"Our aim is not to do away with corporations...On the contrary, these big aggregations are an inevitable development of modern industrialism, and the effort to destroy them would be futile unless accomplished in ways that would work the utmost mischief to the entire body politic...We draw the line against misconduct, not against wealth."

The Gilded Age produced a number of reforms in government-business relations. Regulatory commissions such as the Interstate Commerce Commission (1887), set up to police the railroads, were established.

The *Sherman Act*, passed in 1890, was designed to outlaw contracts and combinations in restraint of trade.

President Woodrow Wilson persuaded Congress to enact the *Clayton Anti-Trust Act,* closing loopholes in Sherman, and to establish the Federal Trade Commission to foster competition and to protect the consumer.

As the shadows of World War I began to lengthen, leaders of American business were becoming more aware that government could be not only friend and promoter, but also a regulator.

***

VII. THE AMERICAN BUSINESS CORPORATION

Economists suggest a variety of reasons for the remarkable development of America's $8.5 trillion economy and the high standard of living its citizens enjoy.

Some say American capitalism, by stressing the profit motive and individual incentive, has contributed to the most productive use of resources.

Others credit a plentiful supply of natural resources, a skilled and educated labor force and an open society. The U.S., with no European feudal tradition, has been the land of opportunity. Free public education has also played an important part.

The American business community, with millions of independent businessmen, has constituted an enormous reservoir of progressive ideas. The nation has enjoyed new and improved products and a rising standard of living.

The existence of the corporation has made some of the nation's economic goals attainable.

Corporations have many of the same legal rights as individuals. A corporation may own, buy and sell property and may sue and can be sued. Some pursue profit, others do not. Corporations may be public or private.

Its charter determines whether the corporation may or may not issue stock. An open corporation offers its stock for sale. Stock in closed corporations is held by relatives, trust funds or various foundations.

A definition of a corporation appeared as early as 1819 when Chief Justice John Marshall called it "an intangible reality" in the famous case of *Dartmouth College v. Woodward.*[26] Marshall called the corporation an artificial but *legal "person"* which can be held responsible for its actions.

Marshall also held that corporate charters were contracts which could not be impaired by state governments. Although states reserved the right to alter, amend or repeal charters, the decision protected private investment from state interference. Coming at a time in U.S. history when capital was relatively scarce, this decision strongly encouraged free enterprise.

The corporation was established so that private individuals could combine their resources for projects too large to be undertaken by other forms of business organization.

Although *sole proprietorships and partnerships* constitute about 80% of all business organizations, corporations, because they include U.S. industrial giants, account for more than two-thirds of all industrial sales.

Modern corporations have developed beyond anything Adam Smith could have imagined. Smith understood an economic system composed of small proprietors, each providing a simple service or product in intense competition with other small businessmen. The corporate form of business organization has brought major changes in American life. Perhaps most important has been the *divorce of ownership and control.*

Stockholders own a part of publicly-held corporations through their shares, but do not ordinarily interfere in day-to-day operations. They delegate control to professional managers and directors.

Corporations are perhaps most attractive because they *limit personal liability* of shareholders for corporate debts. Stockholders can lose only as much as they invest. Corporations can raise considerable sums of capital while stockholders can invest with comparative safety.

U.S. economic history, however, is filled with stories of abuse of power by some titans of industry. The birth of *trusts* and their subsequent demise is one such example of power misused.

In 1879, lawyer Samuel Dodd created the trust as a device for preventing competition in the oil industry. Stockholders of firms wishing to join the Standard Oil Co. Trust were asked to surrender their actual shares to the board of directors of the new trust. In return, they received "trust certificates," giving them the same share in profits as Standard Oil earned.

Directors of Standard Oil exercised control over associated firms while former stockholders received a full share of the profits.

Opposition to trusts developed so rapidly in the late 19th Century that both major political parties denounced them in 1888.

Two years later, Congress passed the Sherman Act, championed by Sen. John Sherman (R., Ohio), which outlawed "every contract, combination in the form of trust or otherwise, or conspiracy in restraint of trade or commerce."

Some critics of corporations claim that they threaten American democracy and destroy the consumer sovereignty Adam Smith visualized. Others attack lack of real competition and criticize *administered prices*, those set by industry itself rather than by competition.

John Kenneth Galbraith, certainly no Conservative, seriously questions some of these attacks on business in his book, *American Capitalism.* Galbraith, setting forth his *theory of "countervailing power,"* suggests that big labor organizations check big business.[27]

Others suggest that new public interest lobbies and government insure that corporate power will be kept within acceptable boundaries.

Galbraith said:[28]

"The size of General Motors is in the service not of monopoly or the economies of scale, but of planning. And for this planning - control of supply, control of demand, provision of capital and minimization of risk - there is no clear upper limit to desirable size. It could be the bigger the better."

***

VIII. THE RISE OF LABOR UNIONS

"Labor Unions are the worst thing that ever struck the Earth, because they take away a man's independence. Financiers are behind the unions, and their object is to kill competition so as to reduce the income of the workers and eventually bring on war...I have always made a better bargain for our men than any outsider could. We have never had to bargain against our men, and we don't expect to begin now."[29]

So spoke Henry Ford in May, 1937, challenging efforts by Walter Reuther and others in the United Auto Workers Union, then attempting to organize Ford workers, as it had done successfully at General Motors and Chrysler.

When Reuther, Richard Frankensteen and others appeared at Ford's Dearborn, Michigan, River Rouge plant to distribute UAW leaflets on May 26, 1937, they were attacked by hired musclemen of the Ford Motor Co., thrown down a flight of concrete stairs and kicked.

From the relatively calm perspective of the late 20th Century, it may be difficult to reconstruct the turbulent history of laboring men and women who fought for the right to join unions and to bargain collectively over wages, hours and working conditions.

Although tensions and industrial conflicts between labor and management appear to be inherent in a free market economy, past anti-labor attitudes and actions have yielded to a calmer process of negotiation and compromise.

Are labor unions really necessary?

"Rugged Individualism"

This question divided workers as well as employers during much of our national history. A deep-seated belief in "rugged individualism" retarded the growth of unionism. So did docile immigrant labor.

But union leaders persisted, arguing that conditions of modern industrial life demanded unions.

"In union there is strength," they said, noting that only through collective efforts could individual workers be protected against the exploitation of long hours, low pay, unsanitary and sometimes dangerous working conditions.

Early efforts to organize workers into labor unions failed. Most employers fiercely opposed unions and rejected collective bargaining. They also resented, in some cases, government efforts to bring labor and management to the bargaining table to resolve disputes perceived to damage the public interest.

During the Great Coal Strike of 1902, George Baer, representing the coal operators, said:[30]

"The rights and interests of the laboring man will be protected and cared for, not by labor agitators, but by the Christian men to whom God in His infinite wisdom, has given the control of the property interests of this country."

President Theodore Roosevelt, intent on ending the strike, summoned Baer and union representatives to the White House. Later Roosevelt is reported to have said:[31]

"If it wasn't for the high office I hold, I would have taken him (Baer) by the seat and the breeches and the nape of the neck and chucked him out of the window."

American courts, applying English precedents, regarded unions as "conspiracies," designed to deprive employers of their legitimate property rights.

The 14th Amendment, ratified in 1868, provided in part, "...nor shall any state deprive any person of life, liberty or property without due process of law."

After courts ruled that corporations were legal persons, entitled to protection of property rights under the 14th Amendment, regulatory legislation directed at certain industries was often struck down or weakened by judicial review.

Courts also commonly used injunctions against unions, effectively breaking up strikes. Ironically, the Sherman Anti-Trust Act, presumed by the public to be aimed at business monopoly, frequently was used against unions.

The strike of the American Railway Union against the Pullman Company in 1894, for example, was enjoined on the grounds that it constituted a conspiracy in restraint of trade under the Sherman Act. Similarly, the U.S. Supreme Court held that a union boycott of a Danbury, Conn., hat manufacturer was illegal and awarded triple damages of $252,000 to the firm.[32]

Unions faced other problems. Unrestricted immigration permitted employers to send abroad for workers. This meant low wages for both the European immigrant and the American worker who competed with him for factory jobs. Henry Clay Frick of Carnegie Steel Corporation and others imported cheap foreign labor. It was a favorite strike-breaking technique.

Historically, most American workers have rejected left-wing, revolutionary unionism and favored "bread and butter" business unionism. The former was represented by groups such as the *Industrial Workers of the World (WWI)*, also known as the "Wobblies," whose Constitution proclaimed:[33]

> "The working class and the employing class have nothing in common...Between these two classes a struggle must go on, until all the tillers come together on the political as well as on the industrial field...Trade unions aid the employing class to mislead the workers into the belief that the working class has interests in common with their employers..."

Some historians regard the *Knights of Labor,* organized in 1869 by a small group of garment workers, as the first successful national American union. Under the leadership of Terrence V. Powderly, the Knights grew to a million members by 1886.

But internal bickering, unsuccessful local strikes, poor leadership and the Chicago *"Haymarket Riot,"* for which the Knights were blamed, combined to destroy the union. Seven persons were killed during the Chicago uprising after a bomb was thrown on May 4, 1886.

The business union approach was taken by Samuel Gompers, who started the *American Federation of Labor (AFL)* in 1881 and led it until his death in 1924. Gompers rejected Marxist notions that workers must rise in rebellion to destroy the capitalist system. He simply wanted a larger slice of the economic pie for working people. Asked to describe his economic philosophy, Gompers once replied: "We want more and more and now!"

Gompers believed in organizing skilled workers - those with a craft. He was not particularly interested in bringing unskilled workers into the AFL. A dissident group which believed that the union's motto ought to be "if it walks, organize it" eventually was to form the CIO.

The leaders of assembly line and other industrial unions vigorously asserted that all workers - skilled and unskilled - should be organized. Only in this way could the AFL meet the workers' needs in modern-day America. Old-line craft union leaders were not convinced. After the 1935 AFL national convention rejected the proposed program for industrial unionism, dissidents took steps to form a new labor organization, the Committee for Industrial Organization (CIO). The union, led by John L. Lewis of the United Mine Workers (UMW), later was to change its name to Congress of Industrial Organizations.

Backbone of the CIO were constituent unions in the following industries: coal, textiles, autos, steel, glass, rubber, oil, gas and refining, among others. In 1937, the CIO unions were expelled from the AFL.

The CIO was considered "radical." Its member unions took on Goodyear Tire & Rubber Company, General Motors, Chrysler and Ford. The United Auto Workers (UAW) led "The Great Sit-Down Strike" at

Fisher Body No. 2 in Flint from December 30, 1936, to February 11, 1937. The dispute paralyzed GM and idled 112,000 of the company's 150,000 workers.

Striking auto workers sung "Solidarity Forever," the union anthem, while their wives, friends and union sympathizers gave them food and clothing through plant windows.

The Flint strike eventually would involve Gov. Frank Murphy of Michigan, President Franklin D. Roosevelt and corporate and union leaders. Murphy called out the Michigan National Guard, not to arrest strikers, but rather to protect them. A Liberal, pro-union Democrat, he feared that a blood-bath would occur if attempts were made to eject the strikers by force. FDR later appointed Murphy to the U.S. Supreme Court.

Ultimately, GM recognized the United Automobile Workers (UAW) as a bargaining agent and agreed not to discriminate against union members. GM further agreed to take up such grievances as the speed-up of the assembly line. By the end of the year all auto firms except Ford had negotiated with the UAW.

The steel industry reacted quickly to the news from Flint. By May, 1937, another key CIO union, the Steel Workers Organizing Committee (SWOC), had won an important victory over U.S. Steel. SWOC claimed more than 300,000 members.

Personalities play an important part in history, and scholars appear to agree that a generation of friction within the labor movement might have been averted had it not been for "personality conflicts."

CIO union leaders such as Walter Reuther of the UAW, Philip Murray of SWOC and particularly John L. Lewis of the United Mine Workers, won important battles for unions. But many leaders of the AFL, who supported labor-management cooperation, disliked the militancy of the CIO. This was particularly true of their leader of the 1930s, William Green.

The AFL and CIO resolved their differences and formed a new united labor front in 1955. Unification was prompted by post-World War II setbacks at the hands of Congress, to be described, as well as a decline in the growth of union membership.

***

IX. UNIONS IN THE MODERN ERA

The election of Franklin D. Roosevelt in 1932 set the stage for dramatic changes in labor-management relations. These fundamental revisions significantly improved the power positions of unions and their members.

Before Roosevelt's New Deal, most U.S. labor law consisted of court decisions. With the support of FDR and the Democratic-controlled Congress, labor won a number of milestone victories.

In 1935, the *Wagner Act* gave workers "the right to self-organization, to form, join, or assist labor organizations, to bargain collectively through representatives of their own choosing, and to engage in concerted activities for the purpose of collective bargaining or other mutual aid or protection."

Employers were forbidden to try to restrain workers in their self-organization and collective bargaining activities, interfere with the operations of any union, discourage workers from joining unions by altering their job conditions or refuse to bargain collectively with elected union representatives.

The *National Labor Relations Board (NLRB)* was created to prevent such interference with the right of labor and to hold representative elections and investigate charges of unfair labor practices. The NLRB could issue "cease and desist" orders where complaints were found to be justified, and seek court support to enforce its decisions.

Roosevelt conceded that the Wagner Act, which imposed no restraints on unions, was one-sided. He felt, however, that it was necessary for government to support unions. Only then, he believed, could labor meet management in our industrial society on anything resembling equal terms.

The President said that the Wagner Act "seeks, for every worker within its scope, that freedom of choice and action which is justly his."

Union leaders applauded the Wagner Act as "Labor's Magna Carta," and some called it "far and away the most significant American labor law ever enacted."

The *Walsh-Healy Act (1936)* provided that companies with Federal supply contracts in excess of $10,000 must observe the eight-hour day and 40-hour work week for employees.

The *Fair Labor Standards Act (1938)* went much further, establishing minimum wages and maximum hours for workers in interstate commerce or those producing goods for interstate commerce. The first Federal minimum wage was 25 cents an hour and the law established the 44-hour week as standard. This later was reduced to 40 hours. The act also effectively outlawed child labor.

The post-World War II years brought a political reaction to labor's growing power and influence. A wave of strikes - in rails, autos, and in the coal mines - antagonized the public. Tired of wartime rationing and controls, consumers wanted to end years of sacrifice and to relax and enjoy the good things of life. When unions struck key industries, they denied these goods to a war-weary American public.

After the Republicans won the mid-term elections of 1946, they teamed with Southern and Conservative Democrats in Congress to enact the *Taft-Hartley Act* over President Truman's veto.

The statute outlawed certain "unfair practices" by unions, limited union status and provided for the handling of "emergency" strikes which threatened the nation's health and safety. Unions were forbidden from coercing workers to become members.

The act forbids *jurisdictional strikes,* which are intramural arguments between labor unions over which of them has the right to a specific job. It also bans *secondary boycotts* - the refusal to handle goods produced by another union - and certain kinds of *sympathy strikes.* The latter are disputes in which unions try to assist another union in gaining employer recognition. Unions may not charge excessive dues. *Featherbedding*, a practice in which union members are paid for work not actually done, was outlawed. Unions may not refuse to bargain in good faith with management.

In the area of union security, the act limited the *"closed shop,"* under which only union members could be hired. It allowed the *"union shop"* under which one need not be a union member at time of employment, but had to join the union within a specific period, usually 30 days.

One of the more controversial sections of Taft-Hartley is 14-B. It allows states to outlaw the union shop and introduced the *"open shop,"* in which union membership is voluntary. Such *"right-to-work"* laws now are on the books in some 20 states which make compulsory union membership and the union shop illegal.

In 1958, the Ohio Manufacturers Association led an attempt to make "right to work" part of the state constitution by popular referendum. It failed and those candidates who embraced the amendment were defeated in perhaps the largest solid labor vote in Ohio political history. Among notable casualties were incumbent Republican Gov. William O'Neal and U.S. Sen. John W. Bricker, considered "unbeatable" by political observers in his race with Democrat Steve Young, then labeled a "non-entity."

Finally, Taft-Hartley outlines a procedure designed to prevent strikes which might disrupt the entire U.S. economy. Under this procedure, the President may obtain an injunction to delay such strikes for an 80-day "cooling off" period. Within that period, striking workers are polled by the NLRB on the employer's "last best offer."

If the offer is rejected, the union then can strike unless the government seizes the industry. This is of questionable legality; witness the U.S. Supreme Court's rejection of President Truman's seizure of the nation's steel mills during the Korean War. *(Youngstown Sheet and Tube v. Sawyer, 343 U.S. 579, 1952).*

The *Landrum-Griffin Act* of 1959 followed a series of hearings before a congressional committee chaired by Sen. John McClellan (D., Ark.) into corruption in labor-management relations.

Disclosure of union corruption, particularly in the Teamsters and Textile Workers Unions, received widespread publicity. Young Robert F. Kennedy, serving as the committee's legal counsel, relentlessly "grilled" Jimmy Hoffa and other Teamsters' officials and established a link between certain union leaders and organized crime.[34]

The Landrum-Griffin Act requires unions to file detailed financial reports, as well as copies of by-laws and constitutions with the Secretary of Labor. It contains "internal democracy" provisions, intended to require free and relatively frequent union elections. The use of secret ballots is an important part of this process. Ex-convicts are barred from union office.

***

X. ORGANIZED LABOR'S GOALS TODAY

What are the major goals and objectives of organized labor on the eve of the 21st Century?

Unions continue to pursue their classic "bread and butter" issues: higher wages, shorter hours and better working conditions. Congress has amended the minimum wage law several times, hiking the rate to $4.25 an hour. It was raised to $5.15 in September, 1997.

Average industrial wages are up sharply. In some industries, automobiles for example, unions have virtually won a guaranteed annual wage. A furloughed auto worker can make up to 95% of his annual wage, whether he/she works or not, under the United Auto Workers contract with major automotive firms.

Union leaders fought hard in the 1970s, but were unable to convince the public of the need for a four-day work week. Some companies are flexible and have unilaterally allowed workers to spend four 10-hour days on the job, rather than the standard five eight-hour days. Nationwide, however, the average work week is only about 37 hours.

Unions have won not only improvements in wages and hours legislation, but also significant changes in *fringe benefits*. The average employee in private industry received more than $3000 in "fringes" in 1994, including vacation, sick leave, pension, profit sharing, health and life insurance, and Social Security payments.

Some union leaders consider the *Employee Retirement Income Security Act,* championed by the late Sen. Philip A. Hart (D., Mich.) one of the most important pieces of social legislation in a generation. The statute makes pensions safer than in the past. Many workers may now keep their pension rights when they change jobs. Perhaps most important, private company pensions are better protected against business failures.

The paramount objective of union leaders today may well be *job security*. They remain actively opposed to accelerated automation and naturally want to protect jobs as modern machines replace workers. This issue, the product of changing technology, will remain a major obstacle in labor-management relations for the balance of the century. Management argues that automation upgrades jobs and even creates new ones. It is difficult to sell that view to longshoremen and union printers, however, as theirs are dying occupations.

One of labor's continuing goals is to improve its *power position* and its influence. It does this in a variety of ways, including campaigning for "friendly" candidates and against those who are "unfriendly," as judged by their positions and votes on issues of concern to labor.

Congress has responded to many of labor's pleas - for *improved minimum wages* and expanded organizing rights over the years. But it has not given it all that it wants. It has failed to repeal Section 14-B of the Taft-Hartley Act, which allows states to bar mandatory union membership. Clinton promised to support repeal in 1992. After nearly 50 years, numerous Democratic Presidents and Congresses have failed to kill the law, leading some to conclude that it may not be as bad as union leaders say.

Labor is teaming up with management, however, to face common economic "threats" from abroad. Both the United Auto Workers Union and the "Big Three" U.S. car manufacturers have a common interest in keeping out Japanese, German and other foreign imports.

George Meany, the late president of the AFL-CIO, warned nearly two decades ago that foreign goods would threaten American producers and American jobs. In a keynote speech to the giant labor federation's convention in Los Angeles, Meany urged President Carter to retaliate duty-for-duty, tariff-for-tariff against countries which trade freely here but bar U.S. imports.

Labor organizations have taken stands on non-economic issues as well. The AFL-CIO backed Carter's Panama Canal agreement. At Carter's request, UAW chief Leonard Woodcock led a U.S. delegation to North Vietnam in the spring of 1977. Virtually all major labor groups enthusiastically endorsed Carter's positions on human rights.

Organized labor has become a powerful political and economic force, although only a minority of American workers are union members. It has, through the process of collective bargaining, won major increases in pay and fringe benefits, shorter hours and improved working conditions for workers.

Of the approximately 130 million men and women in the American civilian labor force today, only about 16% outside agriculture belong to labor organizations. In the private sector, only 11% of the workforce is unionized.

Some students of American labor history note that unions suffered a steady decline in both membership and political clout during the Reagan-Bush Era. Reagan is the only President who also has been a union official. He served some six terms as president of the Screen Actors Guild. But he took a tough anti-union stance when the Professional Air Traffic Controllers Organization (PATCO) went on strike, claiming that understaffing and long working hours were threats to air safety. Technically, the controllers were U.S. civil servants, so the strike was illegal. The President simply fired them for violating their contracts.

In 1982, locomotive engineers struck, curtailing both passenger and freight service across much of the U.S. The President won congressional approval under the Railway Act directing unions to accept an emergency board's proposals for settling the conflict. The unions agreed to comply with the board's findings.

President Bush also was confronted with the prospect of an 11-union national strike of freight railroads. He asked Congress to stop the strike, which it soon did. Bush, however, refused to intervene in an Eastern Airlines dispute, largely on the grounds that people could use other airlines. Eastern subsequently collapsed.

Labor still is strong in large cities of the industrial North, and it has prevailed there in industrial and political conflicts. In the south, southwest and small cities of the midwest, however, unions have been much less of a force. Unions now are centering their renewed efforts at organi-ation in these areas.

Unions will focus future organizational efforts on farm workers, white collar workers and public employees. Mexican-American farm workers in Southern California were caught in the middle of a spectacular battle between the Teamsters and the United Farm Workers (UFW), led by Cesar Chavez.

Rapid growth of teacher unions has been a notable development. Collective bargaining now is a fact of life for many school boards and university campuses where unions once did not exist.

Politically, unions still can be a vital factor in helping to elect Democratic Party presidential candidates, as was demonstrated by the case of Clinton in 1992 and 1996. During his campaigns, Clinton spoke of the need to "empower" labor. He supported the running of companies by management and labor working together.

As President, Clinton invited air traffic controllers fired by Reagan to reapply for their former jobs. He brokered an agreement between striking flight attendants and American Airlines. He chose Robert Reich, a former Harvard professor clearly sympathetic to workers and their unions, as his first Secretary of Labor. The administration supported an increase in the Federal minimum wage and it urged businesses to practice "corporate responsibility."

Unlike Milton Friedman, Clinton and Reich believed that American business has responsibilities to do more than make profits.

Reich told White House reporters in mid-May, 1996, that "...there's still a great deal of insecurity out there among workers who feel that although their companies are doing well, they aren't." [35]

In an address to leading business executives assembled in Washington to honor a dozen companies spotlighted as good corporate citizens, Clinton called on U.S. firms to "do well" by their employees as well as their stockholders. He said: [36]

> "The companies we will hear from are being showcased for one reason. They have done these things in ways that prove you can do the right thing and make money...I believe the power of example to change the behavior of Americans is enormous."

Honored firms and their executives created family-friendly workplaces, offered generous health care and retirement benefits or upgraded worker skills. Some Republicans said that the President, seeking re-election in November, was only resorting to a campaign gimmick and was trying to bash companies enjoying record profits but not sharing them with their workers.

Clinton and the unions have had their differences, notably after the President promoted the *North American Free Trade Agreement (NAFTA)*. The AFL-CIO strongly opposed NAFTA as a threat to American jobs by opening U.S. markets to goods made by cheap Mexican labor. President Lane Kirkland said that Clinton had "stiffed" working people and bribed legislators for votes. But the dispute died down after Congress passed NAFTA. The President needed labor in his bid for re-election in 1996, and the unions needed support for other items on their agenda. If the Clinton-Labor "marriage" was not made in Heaven, it was at least a marriage of convenience.

***

XI. THE RISE OF THE CONGLOMERATE

Some Americans are uneasy about concentration of economic power in large corporations, even though they admire their productive achievements.

"In place of the market system," said economist John Kenneth Galbraith, "we must now assume that for approximately half of all economic output there is a power, or planning system...The planning system consists in the United States of, at the most, 2000 large corporations. In their operation they have power that transcends the market. They rival, where they do not borrow from, the power of the state."

Comparatively few citizens, however, appear to favor radical extensions of anti-trust laws to break up "blue chip" firms merely because they are large.

Americans are divided on this problem. They at once believe in competition's leading to selection of the best, which then becomes the biggest, and in competition, for competition's sake, to prevent concentration of power - monopoly or oligopoly.

"These conundrums can be summarized in saying that the law, as presently interpreted, seems to say that firms should compete, but should not win," writes economist Yale Brozen of the Graduate School of Business at the University of Chicago.[37] "We have confused high concentration with monopoly and competitive activity by large firms with predatory behavior."

Is bigness in modern industry inevitable? Business success stimulates corporate growth, which has tended, in turn, to produce combination.

There have been several "waves" of U.S. business mergers. The first occurred from 1898-1903, followed by another in the late 1920s. Others followed in the late 1960s and 1980s.

In 1968, some 4000 business "marriages" took place, leading some to conclude that merger activity had reached its apex.

With the recovery of the economy from a 1974-75 slump, however, the merger movement again gathered strength. Many large corporations were "heavy" with cash.

United Technologies, manufacturer of aircraft engines, rocket engines and other power plants, sought to take over Babcock & Wilcox in the summer of 1977. B&W was a major manufacturer of boilers for the utility industry. J. Ray McDermott, a relatively small petroleum exploration firm, later bought B&W.

In November, 1977, National Distillers and Chemicals Corp. and Emery Industries agreed to fold Emery into National Distillers for $220 million to Emery shareholders.

Despite merger trends and growth of big business, small businesses also have multiplied. At the dawn of the 20th Century, there were 1.25 million firms in the United States. This increased to 11.5 million by the mid-1960s.

In the conflict between big business and Justice's antitrust division, the limits of the law constantly are being defined and redefined. In 1961, a U.S. District Court found some 29 firms, including Westinghouse and General Electric, guilty of conspiring to fix prices of electrical equipment.

Seven corporate executives were imprisoned, 24 others got suspended jail terms and the companies paid some $42 million in fines.

Some Americans equate mere size with economic concentration. The key question seems to be whether "bigness" in itself is good or bad for the nation's economic growth and well-being.

In 1950, Congress passed the *Cellar-Kefauver Anti-Merger Act* to forbid corporate acquisitions whose effect would be "substantially to lessen competition or tend to create a monopoly."

The first major court test of the new law followed an announcement that Bethlehem Steel and Youngstown Sheet and Tube, the second and sixth largest steel producers, planned to merge. The Justice Department brought them into court in 1958 and the courts prohibited the proposed merger.

Practically forbidden from significant expansion in their own industries, many firms use mergers to diversify. From their perspective, such diversity enables them to counter the rollercoaster drops in the business cycle, to supplement a declining market with a growing one and to take advantage of tax laws by combining tax credits in a losing business with earnings from a profitable one.

Conglomerate mergers are those in which the products of merging firms appear to have no direct or immediate relationship to one another. In 1946, General Tire, confined almost entirely to tire manufacturing, had sales of $106 million. Gradually the company bought interest in a rocket-engine manufacturer and purchased RKO's film library which it quickly recovered in sales of TV rights. In 1963, General Tire was involved in spare products, plastic and chemicals, and sales had passed $1 billion.

***

XII. GOVERNMENT AS REGULATOR

The idea of a state-regulated economy is contrary to traditional free enterprise thought. But a purely competitive economy no longer exists, if it ever did, except in the minds of philosophers.

The U.S. economy is "mixed," characterized primarily by free enterprise but also by a growing role for government.

Municipal government sometimes practices what is called "sewer socialism." That is, public ownership of utility companies. Municipal water works, gas and electric companies and transportation systems are common in the U.S. These are so-called *natural monopolies*.

If not publicly owned, these facilities are regulated by public agencies, empowered to supervise their operations and to approve or disapprove their rates. In return, a municipal utility is protected from competition. Municipal utilities are allowed to exist as monopolies because they must have many customers to spread out the huge cost of laying pipes, stringing wires and building generating stations. Economies of scale are so extensive that one company can supply the market at a lower unit cost than could many competing firms. Clearly, competition would not be in the public interest.

At the national level, public ownership has been established in the cases of the Postal Service, the Tennessee Valley Authority and Amtrak.

The economic role of American government has generally been limited to maintaining conditions under which free enterprise can operate. Government is expected to protect property rights, to establish a sound monetary system, to enforce contracts and to protect national security.

Where exceptions were needed, Congress granted them. The Constitution's authors, particularly Alexander Hamilton, understood the need to stimulate America's economy. Both national and state governments cooperated, for example, to stimulate creation of a transcontinental railroad system. Millions of acres of public lands were granted to rail firms for this purpose.

Over the years, Congress also has aided the trucking industry, enacted laws favoring oil companies and created independent regulatory commissions to protect consumers, but these agencies also limit competition.

Why does government regulate business?

In general, public policy has justified government "intervention" in the free market economy to:
- protect public health, safety, morals and welfare
- maintain equality of opportunity for all citizens
- protect small firms from unfair competition from larger companies
- prevent price fixing and collusion harmful to consumers and small competitors
- conserve natural resources and to protect the environment

The social responsibilities of business are hotly debated in board rooms, university seminars and in the halls of Congress. It may now mean more than living up to high ethical standards and giving generously to charity. Increasingly, it means protection of the environment, truthful advertising and minority hiring, among other things.

Who are the leading regulators? At the Federal level, they include the following:

The *Interstate Commerce Commission (ICC)*. Created in 1887, this agency originally regulated railroads which shipped goods across state boundaries. Its role later was expanded to include trucking, interstate bus lines, water and shipping express companies, among others.

The *Federal Aviation Administration (FAA)* and *National Transportation Safety Board (NTSB)*. These agencies exercise authority over such matters as air safety, routes and tariff rates.

The *Federal Energy Regulatory Commission (FERC)* regulates interstate production and sale of gas and electricity, gas and oil pipelines and water power sites.

The *Federal Communications Commission (FCC)*. The FCC licenses and regulates radio and television broadcasting in the "public interest, convenience and necessity." It may disapprove renewal of a broadcaster's license if he or she fails to meet community responsibilities in programming. However, renewals are rarely denied. The FCC also has jurisdiction over telephones, CB radios and ham operators.

The *Securities & Exchange Commission (SEC)*. A creation of the New Deal in 1934, the SEC was designed to prevent fraud and the kind of reckless speculation in trading of securities that lead in part to the Great Depression. All stocks and bonds offered for sale in interstate commerce or by mail must be registered with this agency so that complete information can be supplied to potential investors.

The *Federal Trade Commission (FTC)*. Created in 1914, the FTC monitors "anti-competitive" practices within the U.S. When the commissioners decide a business is violating laws protecting the consumer's interest, the antitrust division of the Justice Department may sue.

Newspaper headlines tell, for example, of the trial of International Business Machines Co. (IBM) in U.S. District Court in New York, of actions planned to "break up" General Motors Corp. or to prove "monopolistic practices" against American Telephone and Telegraph Co., Western Electric and Bell Telephone Laboratories.

The regulatory role of the U.S. government in the economy has increased dramatically in the past century. It has not destroyed the free enterprise system as some critics have feared. Nor have socialist solutions to the economic problem been attractive to most Americans.

It is clear, however, that business for years was subject to a large and growing list of regulations. Major increases in taxes, federal spending and borrowing all meant an expanded economic role for government. Former President Gerald R. Ford, commenting on Federal regulation of business, once observed:[38]

"Although most of today's regulations affecting business are well-intentioned, their effect, whether designed to protect the environment or the consumer, often does more harm than good. They can stifle the growth in our standard of living and contribute to inflation...Over a period of 90 years, we have erected a massive Federal regulatory structure, encrusted with contradictions, excesses and rules that have outlived any conceivable value."

In recent decades, public officials and economists have concerned themselves with *"externalities,"* such as pollution, problems of occupational health and safety and product safety. These are, some say, the "bads" that inevitably go along with production of economic goods. Concern about social costs in the environment, jobs, energy and the role of government regulation will continue to mark American political debate in the foreseeable future.

***

XIII. GOVERNMENT AS PROMOTER

American government has been not only a regulator of business, but also a promoter of economic progress and development. From the beginning of U.S. history, business has been aided in a number of ways. These include direct aid, such as monetary subsidies, protective tariffs, tax concessions, loans and grants. After the American Revolution, government aided "infant industries" by erecting a high protective tariff wall.

Indirect subsidies also are involved in government promotion of business. The U.S. pays airlines and railroads to deliver the mail. Highway appropriations indirectly aid truckers, bus lines and auto manufacturers. Uncle Sam also finances airport construction.

Tax laws have given credits to business firms to encourage investment in new plants and equipment which, in turn, creates jobs and stimulates community development.

Some historians believe that Franklin D. Roosevelt's New Deal was not only pro-union and pro-farmer, but also pro-business.[39] A wide variety of New Deal agencies spent billions of dollars on economic recovery programs. These funds helped to refinance farms and homes. The U.S. government entered the mortgage field to ease strain on banks, building and loan associations and insurance companies.

Since the 1930s, government aid to the economy has assumed great dimensions.

The Federal government is not alone in promoting economic activity. Most of the 50 states also have economic development agencies or commissions which try to lure new industry through such inducements as tax concessions.

Some states which make few such concessions accuse others of *industrial piracy*. Business interest groups, such as Chambers of Commerce, often work closely with political leaders to promote community economic activity.

When Ford Motor Company announced that it would build a new transmission plant in the Greater Cincinnati area, it was cheered by Lawrence McLaughlin, an economist for the local chamber. McLaughlin said that $140 million in permanent annual income would be added to the area.

"I'm not talking about economics and ripple effects," McLaughlin said. "I'm talking about wages and salaries for the jobs at the Ford plant and new jobs at companies in the area that will supply the Ford plant."

If Cincinnati was happy, Michigan was not. Richard H. Headlee, president of Michigan Taxpayers United for Tax Limitation, said that Ford's decision to build in Ohio emphasized Michigan's need to establish a limit on taxes. "This decision by Ford makes it plain to everyone that Michigan cannot compete with its neighbors for jobs. Given our tax rates, it appears we cannot compete with anyone."

Headlee was to have his day. The Headlee Amendment to the state constitution limits Michigan's authority to "tax and spend." Republican Gov. John Engler subsequently touted his state's ability to compete with Ohio and other states in creating a "pro-business" environment.

The U.S. government is engaged extensively in a large number of promotional activities involving both direct and indirect subsidies to nearly every major economic group.

Agriculture traditionally has received U.S. subsidies. Attempts to solve "the farm problem" have been costly and have generated heated political debate.

In 1862, Congress created the *Department of Agriculture,* an agency for the farmers' interests. In the same year, the Homestead Act gave 160 acres of public land to any settler who filed a claim and stayed on the land for five years.

The *Morrill Act* set aside thousands of acres of public lands for agricultural education.

Government has supported farmers for both social and economic reasons. Historically, the family farm was viewed as the backbone of American democratic life.

The New Deal tried to give the farmer "parity" with other elements of the economy such as business and labor. Overproduction in the 1920s and 30s caused sharp drops in farm income. This led farmers, in turn, to overproduce still more. The *Agriculture Adjustment Act(s)* were intended to prop up farm prices by reducing food supply. Farmers who agreed to limit production got U.S. cash payments.

Over the years, government bought up large farm surpluses. But during the 1970s, market prices were so high that government did not have to intervene for several years. One reason for high prices was the 1972 sale of the U.S. wheat crop to the Soviet government. Prices then soared from $1.70 to $5 a bushel within a year.

Declines in corn prices sparked protests against government agricultural policy during the Carter years and "tractor marches" on Washington, D.C., Carter's hometown of Plains, Ga., and a number of state capitols made headlines.[40] Some farmers apparently followed the advice of 19th Century agricultural protester Mary Lease who advised them, "Raise less corn and more Hell."

Other segments of the economy also have benefited from U.S. aid. Railroads were granted millions of acres of public lands and financial assistance to construct transcontinental lines.

The *Commodity Credit Corporation* and the *Small Business Administration* have extended credit and provided capital to persons who otherwise might not be able to get them.

Since the New Deal, government has given direct aid to financial institutions making mortgage loans. Real estate firms, the construction industry, home appliances manufacturers and other interests all benefit from this.

Many couples have been able to buy homes under programs involving the *Federal Housing Administration (FHA)* and *Veterans Administration.* With the post-World War II population boom, this was a matter of historic importance.

Federal programs also have promoted small business. Between 1932-54, the *Reconstruction Finance Corporation* bailed out many firms from the edge of bankruptcy. The RCFC was replaced by the *Small Business Administration* during the Eisenhower years.

Because politics often is a case of "who gets what," government policies to promote various sectors of the economy are hotly debated. While many citizens criticize specific programs, they appear to support government promotion of economic growth and development.

Indeed, economic stagnation, decline and depression worry many Americans, particularly those old enough to recall the stock market collapse of 1929.

***

XIV. THE GREAT DEPRESSION

Americans, believing that they had found the key to continuing prosperity, were in a highly optimistic mood in the summer of 1929. Industrial production had risen 50% in the past decade and production of goods and services had peaked.

Business profits were high. Labor's increased wages enabled working men and women to buy new autos and household appliances. Some assumed that all one had to do to become rich was to buy stock and put it in a safe deposit box where it could appreciate in value.

John J. Raskob, chairman of the Democratic Party and a director of General Motors Corp., had published an article for the *Ladies Home Journal* entitled, "Everybody Ought To Be Rich."

He wrote:[41]

"If a man saves $15 a week and invests in common stocks, at the end of 20 years he will have at least $80,000 and an income from investments of around $400 a month. He will be rich."

The nation believed prosperity would endure. Republican presidential nominee Herbert Hoover, himself a former successful businessman, said during his 1928 campaign against Democrat Al Smith:[42]

"We in America today are nearer to the final triumph over poverty than ever before in the history of any land. The poorhouse is vanishing from us. We have not yet reached the goal, but given a chance...we shall soon with the help of God be in sight of the day when poverty will be banished from this nation."

As Hoover spoke, the American economy showed great vitality. Only 3.2% of the nation's 49.2 million workers were unemployed. The average work week had declined from 60 hours per week at the turn of the century to 44 hours per week by 1928. Average earnings had increased from 20 cents an hour to 56 cents an hour.

But in the four years between the Great Stock Market Collapse and the election of Franklin D. Roosevelt, the U.S. Gross National Product, a measure of the value of the nation's output, declined 46%, from $104 billion to less than $56 billion.

To the average citizen, the most dramatic development was a sharp rise in unemployment. From virtually a full employment economy in 1919, the U.S. moved to jobless levels beyond any previously experienced. In 1933, some 12.8 million Americans were unemployed, a horrendous 25% of the 51.6 million member labor force. Why?

Economists have tried, for more than 70 years, to explain the Great Depression. It gripped not only the U.S., but the entire world. Although hindsight is often 20-20, economists do not agree, even in 1999, on the exact nature of the economic disease that so afflicted the American people.

Some economists believe that two drags on the economy - agriculture and construction - helped to precipitate the Great Depression. During the 1920s, world food prices declined. Although a boom was underway in most sectors of the economy in the late 1920s, the American farmer was in serious trouble as early as 1926.[43]

The farmer had not achieved parity between the things he had to buy and the products he sold during the entire decade of the 1920s. Mortgage debts increased $2 billion while farm land values declined by $21 billion. Foreclosures and bankruptcy rates soared.

Nature aggravated the farm problem. Drought, dust and despair settled over the Great Plains. Hundreds of "Oakies" (Oklahomans) packed up and headed for California and a fruitless search for jobs. The California legislature passed a law barring the unemployed from entering the state. This statute later was declared unconstitutional by the U.S. Supreme Court.[44]

The Post-World War I business boom was ushered in by a rise in the construction industry. But the construction industry declined between 1925-27 and a serious slump set in by 1928. Investment spending in housing dropped $3 billion between 1929 and 1932.

Heilbroner suggests that maldistribution of income was another important factor in the Great Depression. He notes that:[45]

"The contrast between the top and bottom was extreme. In 1929, 15,000 families with incomes of $100,000 or more actually received as much income as five to six million families at the bottom."

Nearly 80% of the American people had no "discretionary income" at a time when prosperity was based on discretionary buying of consumer durables - autos, furniture, home appliances and housing.

President Hoover cited a sharp drop in U.S. exports as a factor in the Great Depression. Between 1930-31 the *favorable balance of trade (exports over imports)* declined from $1 billion to $500 million. Some economists believe that the 1930 Smoot-Hawley Tariff, which erected a high protective wall against imports, was one of Hoover's greatest blunders. The idea that foreign firms must sell in the U.S. to earn dollars with which to buy America products was not widely understood at the time.

America's banking system also was a key factor in the Great Depression. Since early in the 19th Century, America had small commercial banks which were largely required to provide their own liquid assets. When values fell, these institutions were unable to survive as well as larger and stronger banks.

Between 1923-29, banks were failing at the rate of more than two a day, most of them in small towns. But after the Stock Market Collapse, bank failures accelerated. Some 1400 banks closed in 1932.

***

XV. THE LEGACY OF THE GREAT DEPRESSION

The Great Depression made a deep and lasting imprint on the lives of millions of Americans. It also seriously challenged the traditional view that those who failed in the race for economic success had only themselves to blame,

Economic misery was too widespread to deny that the system had broken down. It was urgently in need of repair if the U.S. was to lift itself out of its worst depression in history.

Newspaper headlines told of unemployed coal miners in Pennsylvania eating wild roots and dandelions. Kentucky miners ate only weeds chewed by cows because they knew that the animals never would eat poisonous vegetation.

In Washington, 10,000 veterans of World War I jammed into tar-papered shanties and waited for Congress to pay their bonuses. When the Senate rejected the bonus bill, Federal troops, under General Douglas MacArthur and his aide, Major Dwight Eisenhower, dispersed the veterans and burned the shanty town.

Radical voices shouted for the destruction of American capitalism. Although most Americans rejected the economic policies of President Hoover, few concluded that the U.S. had to resolve its problems outside the framework of the free enterprise system.

Some, including Hoover, worried that the Federal government would do "lasting damage" to individual character and initiative if it distributed funds to unemployed workers and their families.

He said:

"This is not an issue as to whether people shall go hungry or cold in the United States. It is solely a question of the best method by which hunger and cold shall be prevented...My own conviction is strongly that if we break down this sense of responsibility of individual generosity to individual and mutual self-help in the country...We have not only impaired something infinitely valuable in the life of the American people but have struck at the roots of self-government."

Millions of Americans looked to the new President, Franklin D. Roosevelt, for economic salvation from the Great Depression. Although loved by millions, he was hated by millions of others. Some regarded the aristocratic FDR as a traitor to his class.

The *National Recovery Administration (NRA),* for example, was a favorite target of detractors. NRA, they said, stood for National Run Around.

Some historians contend that FDR could have led the nation in the direction of *Democratic Socialism,* similar to post World War II Britain, had he so desired. But Roosevelt made no such attempt, although some of his critics made such accusations.

Clearly, one of the major legacies of the Great Depression was a more active economic role for the U.S. government.

A group of Harvard and Columbia University intellectuals, called the New Deal *"brain trust,"* accepted the concept of a *"compensatory economy."* Such an economy is a long way from the market system envisaged by Adam Smith in 1776.

Rather than the government keeping its "hands off" the economy, it now sought to create conditions of economic growth and stability. The undesirable effects of the business cycle, with its boom and busts, had to be countered.

For the first time in U.S. history, the Federal government planned to create "economic stability" through strong countercyclical actions. A "sounder capitalism" was the goal of the New Deal. Its objectives were not revolution and radicalism, only recovery and reform.

American capitalism, FDR believed, needed to be revitalized. First, reforms were made in banking, interstate marketing of securities, aid to farmers, workers and businessmen.

The New Deal was long on social and economic experimentation in an effort to get the U.S. economy moving again and to put people back to work. Few Americans today know that FDR pledged in 1932 to balance the budget and to follow other "orthodox" remedies for curing the illness of the Great Depression.

In 1932, FDR talked in glittering generalities. He appeared to offer something to everyone. His San Francisco Commonwealth Club speech, outlining a need for an economic order, appealed to left of center Democrats. But Democratic Party Conservatives, reading the same speech, were pleased that FDR said that Federal regulation of the economy should be a last resort.

In retrospect, one could hardly have predicted the major changes of the New Deal from Roosevelt's campaign rhetoric. He often said that government interference in the economy to achieve economic stability would be kept to a minimum.

He attacked President Hoover for spending deficits and promised to balance the budget. Early in 1933, he sent Congress a message which said in part: "Too often in recent history, liberal governments have been wrecked on rocks of loose fiscal policy."

In short, FDR did not set out to engage in massive Federal spending projects or monumental deficit spending as his legacy to the nation.

Like most other Presidents, FDR was imprecise in his economic views. In his inaugural address, for example, he said that he was a champion of an "adequate and sound currency." When the press asked him to explain this statement, he replied: "I am not going to write a book on it."[46]

***

XVI. THE ROOSEVELT "REVOLUTION"

Some American historians refer to the 1930s as the "Roosevelt Revolution," a period in which the U.S. embarked upon fundamental social and economic changes.

Although some associate revolution with violence and bloodshed, others use the term in a more restricted sense. Libraries are filled with accounts of the "quiet" New Deal Revolution, although there was little quiet about Franklin D. Roosevelt.

Historian Edwin P. Hoyt wrote in his book, *The Tempering Years:*[47]

"...The 1930s were the Roosevelt years, a period in which he dominated American history as George Washington dominated the first 20 years of American nationhood."

FDR's response to the Great Depression was pragmatic, rather than philosophical. Asked to describe his philosophy, the President once replied, "I am a Christian and a Democrat." He refused to go beyond that statement.

Confronted with urgent problems that demanded practical solutions, Roosevelt approved his advisers' experiments. "Let's try it and see what happens," he said. This view shocked and dismayed some of the more Conservative elements in Congress and in his own cabinet.

Soon after assuming office, Roosevelt called Congress into special session. There followed the famous "100 days," in which virtually every FDR recommendation was translated into law. Some statutes were enacted so hastily that they ran afoul of the U.S. Supreme Court. In all, 15 major new laws and scores of minor ones were passed.[48]

Relief was highest priority. Banking got relief through the *Emergency Banking Relief Act*. The farmer needed help and got it from the *Agricultural Adjustment Act,* a Federal Land Bank system, and the *Farm Credit Administration.* American industry got relief from the *National Industrial Recovery Act (NIRA).* FDR and his advisers believed that prices were too low. Therefore each industry was permitted to establish "codes" of fair competition.

Essentially this meant a policy of self-regulation for business as well as temporary suspension of the Sherman Anti-Trust Act. The National Recovery Administration (NRA) reviewed such codes and drew up others for industries which did not produce their own.

The NIRA was intended not only to raise prices, but also to increase wages. Labor got relief with Section 7-A which gave workers the right to join unions and bargain collectively with employers. It also provided minimum wages and maximum hours for workers and eliminated child labor.

In May, 1935, the U.S. Supreme Court found the NIRA to be an unconstitutional delegation of congressional power to the executive branch of government. Labor's legislative goals later were achieved by the *Wagner Act (1935)* and the *Fair Labor Standards Act (1938),* both of which survived Supreme Court challenges.

A major part of the New Deal "Revolution" was relief for the unemployed. If the private sector could not provide work, FDR reasoned, then government should "prime the pump" by becoming the employer.

For jobless youth, there was the *Civilian Conservation Corps,* an agency concerned with land clearance, flood control, reforestation and road building. It employed about 300,000 youths who lived in work camps run by the U.S. Army.

The *Public Works Administration (PWA)* was intended to stimulate the construction industry. Laborers and engineers built roads, bridges, schools and public facilities.

The *Works Progress Administration (WPA)* gave jobs not only to unemployed factory workers, but also to creative artists. WPA theatre created jobs for unemployed actors, writers, painters and stage technicians.

Some New Deal critics, such as General Motors chairman Alfred Sloan, charged that these programs had little effect on recovery, were poorly planned and inefficient. But CCC, PWA and WPA put people to work and gave many Americans a sense of dignity and personal worth that comes with earning one's own way in life.

The Housing Act of 1934 created the Federal Housing Administration. It insured mortgage loans for up to 25 years at low interest rates, generally 5%.

Perhaps the crowning New Deal achievement was the *Social Security Act* of 1935. It created a variety of programs, most familiar of which may be the *Old Age and Survivors Insurance (OASI)* program. Under OASI, insured workers received U.S. pensions on retirement. Widows with dependent children also may claim benefits if a breadwinner dies. Financed by a payroll tax levied on both the worker and employer (originally 1% each), the tax has skyrocketed with increased benefits over the years and has become a major political issue.

Both the political left and right assailed these New Deal reforms, leading Roosevelt adviser Adolf A. Berle to comment, "It is just possible that all of the social inventiveness of the world was not exploded between the poles of Adam Smith and Karl Marx."

These were desperate times for America. The nation could have taken a more extreme path than the New Deal, considering the demagogues of the day.

Huey Long, a Louisiana Democrat, promised to "Make Every Man a King," but his populism was suspect. Indeed, some historians view his administration as the closest thing to a state authoritarian regime in U.S. political history.

Fr. Charles E. Couglin of Royal Oak, Mich., "the radio priest," demanded radical inflation through his National Union for Social Justice. But Couglin, a fierce anti-Semite, ultimately was silenced by the Catholic hierarchy.

Some 103,000 Americans voted in 1932 for William Z. Foster, the Communist Party candidate for President. Others who had lost faith in President Hoover's economic individualism saw a need for a Fascist dictator, like Benito Mussolini in Italy. But efforts to organize an overt, American Fascist political movement were doomed to failure.

Political Scientist James M. Burns, in his book, *Roosevelt: The Lion and the Fox*, wrote that:[49]

"Roosevelt spurned the central concepts of socialization. In 1933, he probably could have won congressional assent to the socialization of both banking and the railroads, but he never tried. He wanted to reform capitalism, not destroy it. And in this sense he was a conservative."

Roosevelt changed American life in many ways. The economic role of the Federal government was enlarged and the "welfare state" was launched. Federal presence was increased in education, labor-management relations and a wide variety of social programs.

XVII. JOHN MAYNARD KEYNES, THE NEW DEAL AND DEFICIT SPENDING

The Great Depression spawned an equally great debate between free-market advocates and economic interventionists over ways to solve the serious economic problems.

The capitalists said that government should let the *natural law of the free market* operate. Forces that triggered the depression would spend themselves and gradually, they believed, natural reversal would occur and recovery would begin.

But political leaders, faced with the enormous magnitude of the Great Depression, considered such a policy politically suicidal and socially inconceivable.

Something had to be done about the jobless who comprised 25% of the 1932 U.S. labor force. These workers and their families could not be permitted to suffer from hunger and malnutrition. Such a short-sighted policy might lead, they feared, to revolution. The extent of human misery and desperation in the land was unprecedented.

Most Americans were not, and are not, familiar with abstract theories of market equilibrium. They did know, however, that they were jobless and hungry.

"People do not eat in the long run. They eat every day," one economic adviser of Franklin D. Roosevelt said.

FDR and his advisers categorically rejected the *laissez-faire* approach. They embraced a new viewpoint, then being widely expounded by the noted English economist, John Maynard Keynes of Cambridge University.

In 1936, Keynes published his best known book, *General Theory of Employment, Interest and Money*.[50] Some considered it the most sweeping revision in capitalist thought since Adam Smith's *Wealth of Nations* in 1776.

Keynes believed that the key to prosperity was the flow of income. He contended that unregulated free markets had failed to provide an adequate flow of income to keep people working. Therefore, he concluded, governments must achieve this goal of social and economic policy.

Keynes rejected the idea that the free market naturally would produce full employment of labor, capital and resources. The market might very well stabilize itself with high levels of unemployed labor, capital and resources.

The *General Theory* was not intended to be a gloomy book, predicting capitalism's doom. Keynes himself was no radical. He accumulated a fortune of more than $2 million by clever dealings in international markets. He believed that most economic activity should be left to free enterprise. He despised Marxist notions.[51]

The cure for the Great Depression, Keynes believed, was greater government spending to make up for inadequate levels of investment by private capital. The Roosevelt administration followed this advice, investing in roads, airports, dams and other public projects, all of which put people to work.

Keynes had another impact on the public economy - the principle of deficit spending. Although FDR came to office promising to balance the budget, he soon found that the programs cost money and that tax revenues were inadequate to finance New Deal policies.

He readily accepted the judgment of his Keynesian economic advisers, most from Harvard and Columbia, who favored a policy of deliberately incurring budget deficits to increase the nation's income flow.

Conservatives considered this sheer economic heresy - which it was at the time. Some businessmen, lacking confidence in the New Deal and its "crackpot Ivy League professors," cut back investments precisely when the United States was trying to "prime the pump."

The debate over deficit financing has become virtually a permanent part of American politics. Some Conservatives believe the nation should balance its budget every year, except in case of war or national emergency. They argue that some state constitutions require such balanced budgets and that these states seem to provide adequate levels of services to their citizens.

Some Liberals argue that the United States need never balance its budget. They observe that state governments have no foreign policies to finance and no large military establishments to support.[52]

Most economists, both in academic and business worlds, seem to reject both extremes. Many feel that budgets should balance over the length of the business cycle, if not every year.

Keynes, a Cambridge don and intellectual, was consistent in his work and theory. He intended his ideas to give a helping hand to enterprise. Clearly, he did not intend to threaten capitalism and capitalists.

Few American politicians really studied or understood Keynes. They departed greatly from his theories. Keynes wanted deficits when needed (recession or depression), balanced budgets in good times and surpluses when the economy boomed.

But members of Congress find it easy to cut taxes and spend when recessions and depressions threaten. They never lose re-election bids by telling constituents how they cut their taxes. Few Presidents and members of Congress, however, are anxious to increase taxes to cope with inflation or hyper-inflation.

In short, what is good economics is not always good politics. And bad economics may be good politics.

Keynes has had such an enormous impact on recent Presidents that one of them, an ardent budget-balancer as a young congressman, remarked: "We are all Keynesians now."

That man was Richard M. Nixon.

***

XVIII. POST WORLD WAR II ECONOMIC PROBLEMS

The Great Depression came to an end with the outbreak of World War II in Europe. During that global conflict, the U.S. had a "full employment" economy. The rate of unemployment for the last three years of World War II averaged only slightly more than 1%. The major problem, in fact, was a shortage of skilled labor in a number of industries.

Many American women, who previously had played the traditional role of housewife, left their homes to work in war plants. "Rosie the Riveter" became a heroine of the domestic front, and some senior citizens of the 1990s can still remember the song of that title. Many American females, having tasted the joys of weekly paychecks and the independence that comes with them, decided to continue working outside the home.

One important legacy of World War II was a significantly greater economic role for the Federal government. Many regarded this as temporary and necessary. Americans were willing to pay practically any price to achieve what was clearly the paramount national goal: victory over the three hated Fascist regimes of Germany, Italy and Japan. Hitler, Mussolini and Tojo were probably the most hated names of the time in the U.S.

Sacrifice was necessary on the homefront, but considering the sacrifices and risks America's fighting men and women were making, there was comparatively little grumbling about *wartime rationing* of consumer goods. Patriotism was a pre-eminent national virtue and the overwhelming number of citizens were willing to do whatever was necessary to support our armed services. Although meat rationing, gasoline rationing and other sacrifices were not popular, all adult Americans understood their purpose.

During the war, Federal spending soared. So did the national debt, which increased from $55 billion in 1941 to $259 billion by fiscal 1945. But inflation was kept somewhat in bounds by wage and price controls, supported by public opinion and patriotism.

The massive task of reconversion from a wartime to a peacetime economy challenged the nation once the military conflict ended. Federal expenditures which had peaked at a yearly rate of $91 billion during the first quarter of 1945, dropped to only $26 billion a year later.

Many Americans were fearful that the postwar era would bring a new recession or depression. But, despite the sharp cutback in U.S. spending, Gross National Product (GNP) declined only about 10% because of capital investment by business firms and a sharp rise in consumer spending.

The immediate problem, in fact, was inflation fueled by excess demand for consumer goods. During World War II, patriotic Americans bought war bonds. They worked long hours and overtime pay was common. Savings mounted. Now, with the end of the war, they were weary of sacrifices. They wanted to cash in their bonds, withdraw their savings and enjoy life.

One student of the period described the nation's mood after V-J (Victory Over Japan) Day as "to hell with duty, let's have some fun."

Consumers wanted an end to wartime controls, such as meat and gasoline rationing. They wanted to be freed from government wartime controls over production plans. Labor unions wanted higher wages, fringe benefits and better working conditions.

In his 1946 budget message, President Truman warned both the Congress and the American people that "inflation is our greatest immediate domestic problem."

But inflationary pressures continued to mount after removal of all wage and price controls, except those on rice, sugar and rents, in November, 1946. Early in 1947, the wholesale and retail price indices both indicated a 30% rise in prices in one year.

Truman urged both labor and management to put the national interest first during collective bargaining procedures and to exercise self-restraint. But unions won substantial wage adjustments and prices soared. Government itself helped to fuel the fires of inflation as the Board of Governors of the Federal Reserve adopted an *"easy money"* policy, making loans available at low interest rates. This encouraged people to borrow and spend and prices of goods already in short supply were bid up still further.

In November, 1947, the President summoned the Republican-controlled 80th Congress into special session and urged it to enact a 10-point anti-inflation program. But the national legislature balked on all except three mild proposals.

The years 1947-48 were characterized by sharp conflicts between the Democratic President and Republican Congress. Twice Congress passed tax cuts in 1947. Both times Truman vetoed them on the grounds that they were the wrong kinds of cuts at the wrong time.

Congress paid little attention to the Chief Executive's warnings about the dangers of inflation and possible recession. In fact, a mild recession began in the fall of 1948 and lasted for about a year.

In an election year atmosphere, the nation suffered through production declines, cutbacks in business investment and falling prices. Congress passed a tax cut of nearly $5 billion over the President's veto.

The 80th Congress also produced the *Taft-Hartley Act*, a measure which its principle Senate sponsor, Robert A. Taft of Ohio, said was designed to restore balance to labor-management relations. The Wagner Act of 1935 had defined a large number of employer practices as unfair. Now, Taft believed, it was time to make unions more responsible.[53]

The Taft-Hartley Act was, in large measure, a result of growing public discontent with strikes. By fall, 1945, a wave of postwar labor disputes erupted. Some 28,000 coal miners were idle. Walkouts occurred in both lumber and petroleum industries. The United Auto Workers struck General Motors, idling 175,000 workers in 19 states. Truman believed that these strikes were understandable, simply labor's readjustments from a wartime to a peacetime economy. As a New Deal-Fair Deal Democrat, he was "friendly" to unions, he said.

But when the railway unions called out 300,000 workers on strike in May, 1946, a national transportation emergency occurred. The President persuaded 18 of 20 rail unions to accept a compromise settlement of their dispute with management. But the Brotherhoods of Railway Trainmen and Locomotive Engineers adamantly refused to go along.

An irate President then went before Congress and requested authority to draft the strikers into the armed forces. Before he had finished reading his speech, Truman was handed a note announcing that the rail strike had been settled on terms recommended by the Chief Executive.

In May, 1946, the U.S. seized the nation's coal mines after they had been closed down for 45 days at a cost of 90 million tons of coal. When United Mine Workers President John L. Lewis refused to comply with a Federal court injunction barring the strike, he was fined $10,000 and his union some $3.5 million. At that point, Lewis ordered the miners to return to work.

But the voters were angry. They elected a Republican Congress in November, the first since 1930. It produced Taft-Hartley, which gave Truman a chance to restore his political position with union leaders. Angered by his actions against rail and mine workers, some union leaders called Truman "a political accident," a reference to his accession to the Oval Office on the death of Franklin D. Roosevelt.

But Truman's veto of Taft-Hartley, although overridden by the 80th Congress, helped him to win labor support in his 1948 upset election victory over the Republican governor of New York, Thomas E. Dewey.

By June, 1950, when the North Korean Communists invaded the Republic of Korea, the economy was on the upturn. Following outbreak of hostilities in Korea, a wartime boom occurred. Many consumers, fearful of a return to World War II type rationing, began to hoard goods they believed would be scarce. Congress increased taxes by $9 billion. Controls on credit were tightened.

As is usual during wartime, Congress granted the President special powers to deal with the emergency. The *Defense Production Act* gave Mr. Truman broad authority to regulate the American economy. He was authorized to allocate materials and facilities, requisition property for defense production and assign production priorities.

He also was authorized to impose controls on consumer credit, to make and guarantee loans to bolster war production and to stockpile scarce materials through a program of long-term supply contracts. The President was also empowered to impose controls over wages and prices.

But Korea was not a declared war. Consequently, when the Chief Executive seized the steel mills during a labor-management dispute, the action was contested. The U.S. Supreme Court held that the President had exceeded his authority and ordered the mills returned to private ownership. The President, the court said, should have used the Taft-Hartley Act, the measure that Congress had provided for emergency strikes. For clearly political reasons, Truman had rejected this course of action.

Truman spent the balance of his term trying to end the long and increasingly unpopular stalemate in Korea. Economic issues had to take a back seat to military, national security and foreign policy concerns.

Many historians, looking back at the Truman administration, consider employment security one of its major accomplishments. Americans still were debating the extent to which government should assume responsibility for providing jobs on the eve of the 1996 election.[54]

***

XIX. THE EMPLOYMENT ACT OF 1946

The Employment Act of 1946 was one of the legislative milestones of the Truman Administration.

Congress, fearing a post-war recession, reaffirmed the New Deal commitment that government would play a major role in the U.S. economy.

The Act declared, "...it is the continuing policy and responsibility of the Federal Government...to promote maximum employment, production and purchasing power."

The idea of full employment legislation was developed in the United States in 1942 when the *National Resources Planning Board* recommended "a positive program of postwar economic expansion and full employment, boldly conceived and vigorously pursued."

A year later, the board urged government to commit the country to a policy of full employment. Government should, the board said, "promote and maintain a high level of national production and consumption by all appropriate measures."

The National Resources Planning Board view was consistent with John Maynard Keynes' *theory of compensatory fiscal policy*. But some economic Conservatives rejected such a policy. To them it was equivalent to accepting permanent, planned budgetary deficits.

Such deficits, they reasoned, could trigger disastrous inflation and lead the nation down the path to financial ruin.

But in the presidential campaign of 1944, both President Roosevelt and New York Gov. Thomas E. Dewey, the Republican Party nominee, endorsed the NRPB view.

Roosevelt, speaking of an American "Economic Bill of Rights," argued that people had a right to a useful and remunerative job.

Governor Dewey said:

"If at any time there are not sufficient jobs in private employment to go around, the government can and must create job opportunities because there must be jobs for all in this country."

Sen. Robert A. Taft of Ohio, then called "Mr. Republican" by many of his admirers, objected to the original full employment bill, introduced in early 1945. He said:

"I am in favor of economic planning but the theory of this bill means a steady expansion of Federal power. It provides immediate recourse to Federal spending."

Taft viewed such a precedent as unwise. It was, he said, like taking a dangerous drug. "If we once begin to take it, we never can escape it," he said.

Taft, concerned about a balanced budget and fiscal responsibility, offered an amendment to the bill, which stated:[55]

"...any program of Federal investment and expenditure for the fiscal year 1948, or any subsequent fiscal year when the nation is at peace, shall be accompanied by a program of taxation over a period comprising the year in question and a reasonable number of years thereafter, designed and calculated to prevent during that year any net increase in the national debt...without interfering with the goal of full employment."

Taft was persuasive, the Senate voting, 82-0, to accept his amendment. But the House of Representatives substantially rewrote the Senate bill. Emanuel Cellar (D., N.Y.) attacked the Senate measure as a document "written by the best minds of the 18th Century."

The legislation was carried over until the next session of Congress when it finally was enacted. In signing the measure, President Truman called it "not the end of the road, but rather the beginning...a commitment to take any and all of the measures necessary for a healthy economy."

The law opens with a philosophical declaration:

"The Congress hereby declares that it is the continuing policy and responsibility of the Federal government to use all practicable means consistent with its needs and obligations and other essential considerations of policy, with the assistance and co-operation of industry, agriculture, labor and state and local governments to coordinate and utilize all its plans, functions and resources: (1) for the purpose of creating and maintaining in a manner calculated to foster and promote free competitive enterprise and general welfare, conditions under which there will be afforded useful employment opportunities, including self-employment, for those able, willing and seeking to work, and (2) to promote maximum employment, production and purchasing power."

The Act created a *Council of Economic Advisers* to "assist and advise" the President in the preparation of a required yearly economic report to the nation. It also created a *Joint Economic Committee* of Congress, enabling the legislature to formulate its own ideals on the state of the nation's economic health.

Some Liberals, noting the word "full" had been removed from the act, were unhappy with it. But others accepted the idea that the government could not guarantee individuals a job.

Taft assured Conservatives "there is nothing in the bill to which any member should take exception."

In the half century since the Employment Act was passed, U.S. policies have fostered expansion of jobs and production. The labor force has more than doubled, from 61 million in 1946 to nearly 140 million in 1999. Unemployment, however, remains a continued source of national concern.

The U.S. has suffered from a number of mild recessions since 1946. Technically a *recession* occurs when there is a drop in total production, income, employment and trade for six months or longer. Many economists, including Heilbroner, describe depressions as nothing more than long lasting recessions.

In the presidential election of 1980, Ronald Reagan quoted an old line: "When your neighbor loses his job, it's a recession, when you lose your job, it's a depression." He added: "When Jimmy Carter loses his, it's recovery."

Declines occurred in 1948-49, 1953-54, 1957-58, 1960-61, 1974-75, 1980-82 and 1990-91. The 1974-75 slump was the most serious since the Great Depression. Although mild, the recession of 1990-91 cost President Bush his job.

Economists often define *"full employment"* in the U.S. as a rate of 4% unemployed or less. Since Congress passed the Employment Act of 1946, the yearly unemployment average has seldom been below that figure. We have not attained a 4% rate in more than a quarter of a century. Only in 1948, 1952, 1966-67-68-69 did we meet the law's target. Today some economists believe that a 4% unemployment rate is impractical, given changes in the composition of the labor force.

A vast influx of women and teenagers into the job market has complicated matters. These groups traditionally have had hard times getting and holding jobs. The role of the U.S. government in providing jobs was hotly debated during the campaign of 1996, and appears likely to be argued in the years ahead.

***

XX. ECONOMIC POLICY DURING THE EISENHOWER YEARS

When Dwight D. Eisenhower was inaugurated in January, 1953 - the first Republican President since Herbert Hoover - some believed the United States had left behind the New Deal-Fair Deal for a new era. Republican control of both houses of Congress tended to support this view.

In the field of economic policy, Eisenhower promised "security with solvency." A military hero and symbol of Allied victory in Europe in World War II, Eisenhower was able to persuade Congress to cut military spending. Congress, certain the President would not jeopardize national security, went along.

Eisenhower saw inflation as the major danger to the U.S. economy. He promised to work toward a balanced budget and to restore fiscal integrity. He also sought an end to the Korean War and imposed wage and price controls, which he called "unworkable and unsatisfactory."

The President greatly admired successful businessmen. Some reporters cynically called his first cabinet "nine millionaires and a plumber." The latter referred to Labor Secretary Martin V. Durkin, a former plumbers' union official. Durkin resigned from the cabinet when the President refused to support his proposals to amend the Taft-Hartley Act.

Eisenhower stressed private incentives, free enterprise and economic freedom. He believed in local initiative and was highly skeptical of government trying to solve people's problems.

Soon after assuming office, Eisenhower faced a mild recession. A promptly enacted tax cut stimulated consumer spending and recovery was underway by the last quarter of 1954. Democratic Party orators, however, reminded voters that the GOP was the party of the "Hoover Depression" and hard times and Republicans lost control of Congress in the mid-term elections.

Eisenhower's first term ended with the nation enjoying relative prosperity. The GOP campaign theme in 1956 was "peace and prosperity." The general, it was claimed, had led the United States out of the Korean stalemate - away from fiscal irresponsibility, deficit spending and excessive government control over their lives. He won a landslide re-election victory, again defeating his 1952 Democratic Party opponent, former Illinois Gov. Adlai E. Stevenson.

Early in his second term, Eisenhower again warned of the perils of inflation. He urged both unions and management to place national interest first in collective bargaining. An unending wage-price spiral must be avoided, he said. Promising that his administration would do its utmost to protect the value of the dollar, Eisenhower said in January, 1957:[56]

> "Business in its pricing policies should avoid unnecessary price increases, especially at a time like the present when demand in so many areas presses hard on short supplies...Increases in wages and other labor benefits negotiated by labor and management must be reasonably related to improvements in productivity...Except where necessary to correct obvious injustices, wage increases that outrun productivity are an inflationary factor."

A nine-month recession that began in 1957 stimulated demands for government job creation policies. Eisenhower refused, saying he was opposed to "going too far with trying to fool with the economy."

Some Republican congressmen, facing re-election in 1958, also demanded action when unemployment reached 4.6 million. But the President preferred a "go-slow" policy. He denounced Federal job creation to counter recession.

The Democrats won the congressional elections and promptly tried to pass a number of programs to stimulate the economy. Eisenhower, however, denounced those who pushed such measures as "budget busters," and wielded his veto.

Despite restraints, the President saw what was then the largest peacetime deficit in U.S. history and the highest rate of unemployment (6.8%) since the end of World War II. One major disruptive factor was a 116-day steel strike.

Eisenhower urged Congress to amend the Employment Act of 1946 to make price stability a national economic goal. The proposal was strongly backed by business, but died in Congress where labor opposition was effective. Unions feared such a provision would conflict with the law's full employment goals.

The economy improved in the last half of 1959 and first half of 1960. But a fourth postwar recession began in May, 1960. Nearly four million workers, 5.5% of the civilian labor force, were unemployed.

This was to the distinct political advantage of Sen. John F. Kennedy of Massachusetts. In his televised debates with Vice President Richard M. Nixon, Kennedy spoke of the need to get the United States "moving ahead again." This was a simple idea, easily grasped by voters, many of whom were suffering from "hard times."

Kennedy said: [57]

"This is a prosperous country. But it can be the most prosperous country in the world...we have to grow, and under Republican leadership, this country is standing still...With a really healthy rate of growth, this country can have a full employment for all who want a job. With a really healthy rate of growth, we can pay for all the defenses that this administration says we can't afford. With a really healthy rate of growth, we can afford the best schools for our children and the best paid and best trained teachers, and finally, with a really healthy rate of growth, we can talk about an economic crusade for justice. But it's time we stopped talking about it and elected an administration that will do something about it."

Nixon responded:[58]

"The charge that the United States has been standing still just doesn't hold up. What they are looking at when they make that charge is solely at what government does...they overlook the very great fact...that America's progress comes primarily not from what the government does, but from what 180 million free, individual Americans do in individual enterprise...these proposals Senator Kennedy has made will result in one of two things. Either he has to raise taxes or he has to unbalance the budget. If he unbalances the budget, that means you have inflation, and that will be, of course, a very cruel blow to the very people, the older people, that we have been talking about...I believe basically that what we have to do is to stimulate the sector of America, the private enterprise sector of the economy, in which there is the greatest possibility for expansion."

Nixon spoke of the Gross National Product, the value of all of the nation's output of goods and services. He compared U.S. and Soviet bases for growth and said that the U.S. was far ahead of the Russians.

His point that it is easy to register spectacular statistical improvements in production when your economy has a smaller base, was, however, lost on many Americans, untutored in the "dismal science" of economics. As often happens in presidential campaigns, the candidate who communicates wins. Nixon's point, while valid, was too sophisticated. The idea that the U.S. needed to "move ahead" was apparent, and millions of voters found Kennedy's arguments more convincing.

***

XXI. KENNEDY'S "NEW FRONTIER" ECONOMICS

On the eve of President-elect John F. Kennedy's inauguration, one of America's favorite pastimes was "naming the cabinet." Rumors swept the nation. One, which horrified Conservatives, was that the young President was going to select John Kenneth Galbraith, Harvard economist and a leader of the left wing intellectuals of the Democratic Party, as Secretary of the Treasury.

They greeted the nomination of Professor Galbraith as U.S. ambassador to India and that of Republican investment banker Douglas Dillon as Secretary of the Treasury with great relief. Perhaps Harvard was not taking over the U.S. government after all.

Kennedy named Walter Heller, head of the economics department at University of Minnesota, as chairman of the Council of Economic Advisers.

Heller, originally introduced to Kennedy by Sen. Hubert H. Humphrey, was no wild-eyed radical. Many of his ideas were old-style New Deal.

Heller was not alarmed by prospects of budget deficits. He considered them normal, perhaps even useful during recessions, to stimulate the economy. But, like Keynes, Heller believed that budgets should be balanced over the length of the business cycle and that surpluses should be achieved during inflationary periods.

Heller and Kennedy were concerned primarily not with inflation, but rather with the recession - inherited from Eisenhower. Although the slump "bottomed out" soon after Kennedy took office, unemployment was more than 8% in February, 1961. A year later, it was down to 6.1%, but this was still a rate considered "socially unacceptable" by Kennedy and his advisers.

Kennedy, in an effort to win support from the business community, proposed a 7% investment tax credit. Such an incentive, he believed, would stimulate capital goods industries and spread production and jobs throughout the economy by what economists call a *"multiplier effect."*

Congress enacted the *investment tax credit,* but refused to give the President power to cut personal income taxes during periods of national emergency. Such presidential actions, inevitably reducing tax revenues, should be accompanied, congressmen warned, by corresponding reductions in U.S. spending.

The investment tax credit later was abolished at President Johnson's request, but was re-enacted by Congress and increased during the Nixon Administration.

Over time, Congress has demonstrated its willingness to cut taxes and, simultaneously, to increase Federal spending.

Traditional economics assumes that business cycles are normal. It also holds that economic fluctuations can be controlled, but not eliminated entirely. Heller thought otherwise. He was a leading disciple of the *"New Economics."*

The "New Economics" assumes that the U.S. economy has an "ever rising" potential. After great effort, Heller converted Kennedy to this position. Kennedy then began to speak of the gap between *"actual"* and *"potential" Gross National Product* - what we were producing and what we could produce in a full employment economy. Kennedy told Congress that the United States was operating at a level between $20-$25 billion below its potential.

In a major speech on December 14, 1962, Kennedy told the Economics Club in New York:[59]

"Our true choice is not between tax reductions...and the avoidance of large Federal deficits...so long as our national security needs keep rising, an economy hampered by restrictive tax rates will never produce enough revenues to balance our budget, just as it will never produce enough jobs or enough profits. It is a paradoxical truth that tax rates are too high today and tax revenues are too low and the soundest way to raise revenues in the long run is to cut rates now...Our practical choice is now between two kinds of deficits: a chronic deficit of inertia...or a temporary deficit of transition..a future budget surplus. The first type of deficit is a sign of waste and weakness. The second reflects an investment in the future."

Early in 1963, Kennedy asked Congress for an across the board $13.6 billion tax reduction. Congress responded slowly, In September, 1963, slightly more than two months before his assassination, the President went on television and pleaded the urgency of tax reduction:[60]

"This is a matter which affects our country and its future. We are talking about more jobs...about the future of our country, about its strength and growth, and ability as the leader of the free world. We are talking...about one of the most important pieces of legislation to come before Congress this year, the most important domestic economic measure to come before Congress in 15 years."

What the President was proposing was revolutionary then, but the "New Economics" - the idea that the United States deliberately incur a budgetary deficit at the peak of the business cycle in order to counter unemployment and economic stagnation - became the conventional wisdom of the late 1970s. Rep. Clarence Cannon (D., Mo.), chairman of the House Appropriations Committee, reacted with scorn:

> "At the heart of our national finances is a simple inescapable fact, easily grasped by anyone. It is that our government...like individuals and families, cannot spend and continue to spend more than they take in without inviting disaster."

Former President Eisenhower asked:

> "What can those people in Washington be thinking about? Why would they deliberately do this to our country? The deficit road throughout history has lured nations to financial misery and economic disaster."

Although President Kennedy never lived to see his tax reduction passed, Congress did enact it at President Johnson's prodding.
Johnson said that the tax cut of 1964 was "...the single most important step that we have taken to strengthen the economy since World War II."
Walter Heller and his "New Economics" school won the day. The U.S. economy responded to the tax cut and unemployment fell. But then Johnson sent the U.S. armed forces into the Vietnam War and his effort to provide Americans with both "guns and butter" planted the seeds of the raging inflation of the 1970s.

***

XXII. LBJ'S WAR ON POVERTY

Although most Americans tend to associate the "War on Poverty" with the Johnson Administration, it actually began under President Kennedy.
Kennedy was distressed by the poverty he saw while campaigning in the West Virginia Democratic presidential primary of 1960.
Theodore White, in *The Making of the President, 1960,* writes:[61]

> "He could scarcely bring himself to believe that human beings were forced to eat and live on these cans of dry relief rations...Of all the emotional experiences of the pre-convention campaign, Kennedy's exposure to the misery of the mining fields probably changed him most as a man."

Kennedy instructed Heller to study the poverty problem in detail. He then urged Congress to go beyond existing Federal welfare programs.
But it remained for Johnson to get the congressional machinery moving. In his first State of the Union Address, Johnson declared war on poverty: "This administration today, here and now, declares unconditional war on poverty in America. It will not be a short or easy struggle, but it is a war we cannot afford to lose."
LBJ signed the *Economic Opportunity Act* into law in August, 1964. The act created an *Office of Economic Opportunity (OEO)* within the President's Executive Office. OEO was headed initially by Sargent Shriver, JFK's brother-in-law, who later was chosen as 1972 Democratic nominee George McGovern's

running mate. This followed the disruptive "Eagleton affair" in which McGovern's first choice for vice-president was dumped from the ticket when the media revealed his history of mental illness. Today's collegians may recognize Shriver more for being Maria Shriver's father.

OEO, intended to coordinate spending of several other Federal agencies, also operated many of its own programs. Each of these was directed toward a specific group of poor people or some particular cause of their poverty.

Many anti-poverty programs, such as the *Job Corps,* stressed education and manpower training. Youths from bad home situations were offered remedial education as well as vocational training opportunities.

But the Job Corps ran into serious problems, not the least of which was cost. The bill for keeping a youth in the Job Corps ran to $7500 a year.

Many camps were located in rural areas and residents were annoyed by the behavior of incorrigible ghetto youths. Many youngsters also dropped out of the program.

An innovative and controversial project was the *Community Action Program (CAP).* Federal funds were channeled to local communities to involve "maximum feasible participation" of residents, the intended beneficiaries of CAP.

In retrospect, program guidelines were much too vague. OEO and the Bureau of the Budget (now Office of Management and Budget) debated aspects of planning. Big city mayors, such as the late and legendary Richard J. Daley of Chicago, were happy for the poor to receive aid - providing their own Democratic Party organizations determined how Federal funds were to be spent.

A number of CAPS came under attack. In Newark, N.J., for example, leaders were blamed for inciting riots in 1967. In New York, Puerto Ricans and Blacks fought for control of the program. Ultimately, local politicians won out, but President Johnson himself cooled on CAP.

LBJ was sensitive to criticism that U.S. funds were spent carelessly, that fraud and corruption marred the program, and that the poor did not benefit much from CAP.

"Operation Headstart" was a favorite of Congress. Pre-school children received early education before becoming "disabled" by poor home environments. The theory behind "Headstart" was that later, at age five or six, children would receive greater benefit from formal classroom instruction.

The *Neighborhood Youth Corps (NYC)* was intended to help unemployed teenager dropouts, or those on the verge of quitting school. Youngsters were given jobs after school and during vacations. The program kept jobless teens off the streets. As a result it never was unpopular, like the Job Corps.

Liberals and Conservatives debated the "War on Poverty" during the late 1960s and early 1970s. Some Liberals complained that the program was only a "Skirmish on Poverty" due to inadequate funding.

Some Conservatives argued that the whole program was a "ripoff" of the taxpayers, intended to benefit Lyndon Johnson politically rather than uplift the poor. Others said the whole program left the basic causes of American poverty untouched.

Johnson, of course, disagreed. In *The Vantage Point,* written after he retired, LBJ said:[62]

"When I left office, government reports showed that of the 35 million Americans who had been trapped in poverty in 1964, 12.5 million had been lifted out...Not only because of the War on Poverty, but also because of the expanding economy, people were coming out of poverty...at a rate two and a half times faster than at any time in our history."

To a large extent, the War on Poverty was the victim of another war - Vietnam. Near the end of his administration, Johnson finally concluded that the U.S. could not have both "guns and butter."

***

XXIII. LBJ, VIETNAM AND THE U.S. ECONOMY

President Johnson's dreams of a "Great Society" ended in an economic nightmare for the American people. Inflation and hyperinflation took roots in the escalation of the Vietnam War. They came to full flower in the next decade.

The erosion of the value of the American dollar was spectacular. The real value of American money declined more than 25% between 1967 and August, 1974, when Richard M. Nixon resigned the presidency.

Gross National Product (GNP), a measure of the value of goods and services produced by the nation's economy, doubled between 1965-75. But most of that gain was an illusion. Real gains in production, after making allowances for inflation, were only about 33% in 10 years.

Many economists, Democrats and Republicans alike, believe that the seeds of the nation's economic tribulations were planted in 1966. Johnson escalated America's role in Vietnam without asking Congress for new taxes to pay for the expanded military effort.

On the economic front, the middle and late 1960s were characterized by rapidly growing defense budgets, high capital investment by business and high consumer demands. Economic pressures thus generated by government, business and consumers fueled an inflationary blaze that eventually roared out of control.

Initially, Johnson apparently believed that the United States could have both "guns and butter." LBJ pushed *"Great Society"* programs at home while the Vietnam conflict became more and more costly in terms of both blood and treasure.

Liberal, as well as Conservative, economists asked if Johnson's program of increased Federal spending had not pushed demands for goods and services beyond the nation's ability to produce them. Scarcity, as any economist knows, leads to rising prices.

The President finally decided in 1967-68 to "dampen down" the fires of inflation by restraining the economy. LBJ expressed willingness to cut "nonessential" Federal spending at home to finance the increased cost of the war.

Seeking to "cool down" the economy, he urged Congress to suspend the 7% tax credit given business earlier to stimulate capital spending. Congress complied, but it was not until August, 1967, that Mr. Johnson urged a 10% surcharge tax on both individual and corporate incomes.

The Chief Executive was slow to ask for higher taxes. But Congress was even slower to enact them. It delayed action until the summer of 1968, when some economists concluded it was too late to have the intended result. But the 10% income-tax surcharge was joined with a $6 billion cut in Federal spending.

Use of *fiscal policy* (taxes, budget and spending levels) is only one means government has to promote economic stability.

A second tool is *monetary policy*. This involves management of the U.S. money supply by the Federal Reserve Board to ensure credit in quantities and at prices in harmony with U.S. economic policy goals.

In 1966, monetary policy sought drastically to reduce the rate of growth of the money supply. This caused interest rates to soar to their highest levels since the 1920s.

All Americans were trapped by a *"tight money"* policy. Borrowing was extremely costly. The building industry in particular was hit hard. Many prospective home buyers could not get loans.

Congress, responding to well-organized political pressures, finally enacted legislation imposing interest ceilings on certain types of loans. It also pumped extra money into the housing industry itself.

By the end of the Johnson years (January, 1969), the U.S. economy was tottering. Prices not only were rising at the rate of more than 5% yearly, but the nation also was suffering from growing unemployment - then at the rate of about 6%.

Traditional textbook economics taught that government can take certain corrective measures to combat inflation or unemployment. There was, most economists believed, an inevitable "tradeoff" between inflation and unemployment. Perhaps some increase in unemployment was the price that had to be paid for price stability.

A political decision was required. Either pay the social price of higher joblessness for price stability or incur persistent and unacceptable price rises for all Americans by making full employment the paramount goal.

But economists had not seriously suggested that unemployment and rising prices would occur simultaneously. To their dismay, the science of economics and particularly economic forecasting came into growing disrepute.

A new word, *"stagflation,"* was coined in Britain to describe this combination of inflation and a stagnant economy. It was a word few Americans, except professional economists, understood.

***

XXIV. NIXONOMICS

When Richard M. Nixon became President, he considered inflation to be the greatest domestic problem facing the United States.

Nixon believed that President Johnson had planted the seeds of inflation in 1965 by escalating the Vietnam War without asking Congress for tax increases to pay for it.

Rapidly rising prices deeply troubled American consumers. Nixon was willing to risk temporarily higher levels of unemployment to halt inflation. Price stability was the paramount goal.

But Nixon remembered that President Eisenhower had suffered through three recessions while trying to curb inflation. This was politically disastrous to the Republican Party. Democrats scored major victories in the 1958 congressional races, then elected John Kennedy as President in 1960. Perhaps most important, the Eisenhower recessions reinforced the old image, planted during the Great Depression, of the GOP as the party of "Hard Times."

In a television address during the 1968 campaign, Nixon said:

"Inflation penalizes thrift and encourages speculation. Because it is a national and perverse force, dramatically affecting individuals, but beyond their power to influence, inflation is a source of frustration for all who lack greater economic powers."

Liberals were angered by Nixon's claim that inflation was the nation's most important domestic problem. Nixon was less than enthusiastic about other problems such as civil rights and about Liberal programs for government to "improve society."

Nixon appeared to have no real technical interest in or knowledge of economics. He left economic policy largely in the hands of his advisers. These originally included Paul McCracken, chairman of the Council of Economic Advisers, Budget Director Robert Mayo, Treasury Secretary David Kennedy and William McChesney Martin, chairman of the Federal Reserve Board.

In their book, *Nixon in the White House,* columnists Rowland Evans Jr. and Robert D. Novak, comment:[63]

"McCracken was the essence of the pedagogue...As McCracken explored every conceivable side of his subject, Nixon's eyes would glaze over."

What policy would the administration follow to fight inflation and seek price stability? Initially, Nixon rejected outright Federal controls on both philosophical and pragmatic grounds. Without rationing and the kind of patriotism the nation showed during World War II, controls wouldn't work.

The original Nixon "game plan" was to hold down the growth of the money supply while trying to achieve a budget surplus. Clearly, the President wanted no $28 billion deficit, like that of LBJ's last year in office.

Monetary policy was no immediate problem. The Federal Reserve Board already had started a "tight-money" policy. But fiscal policy - use of tax, budget and spending policies to stabilize the economy - was another matter.

The 10% income-tax surcharge, which took effect in 1968, was due to expire. Arthur Burns, a Columbia University economist who was an expert on the business cycle of growth and recession, and others feared the loss of $10 billion a year in revenue. They considered it folly to permit the surcharge tax to lapse.

A compromise was struck. Congress extended the surcharge for a year at the rate of 5%, after which it was permitted to die.

Nixon cut $4 billion from LBJ's final budget and confidentially predicted a surplus of nearly $6 billion. By summer, 1969, Nixon's early policy, called *"gradualism,"* was in deep trouble.

The President then turned to *wage-price guidelines*. Also, early in 1970, Nixon nominated Burns to replace Martin as Federal Reserve chairman.

Under Burns, the Fed scrapped the "tight money" policy. Several key economic indicators were down. Industrial production was down. So was personal income. For the first time since World War II, real Gross National Product (GNP) declined. Stocks slumped.

But consumer prices and interest rates both were up. This bit of bad economic news caused Lawrence O'Brien, then chairman of the Democratic National Committee, to invent the term "Nixonomics."

"Nixonomics," O'Brien said, "means that all things that should go up - the stock market, corporate profits, real spendable income - go down, and all the things that should go down - unemployment, prices and interest rates - go up."

In December, 1970, Nixon promised a "bare bones" budget to stop inflation. But the deficit compared to Johnson's worst.

By August, 1971, the situation was desperate. Acting more vigorously, Nixon imposed a 90-day *freeze on wages and prices* (Phase I) during which time Federal Pay and Price Boards were established to regulate key sectors of the economy in Phase II.

The controls held inflation down to less than 4% a year. The economy surged forward late in 1972 and early in 1973.

Burns pumped out new money to fuel the economy and help Nixon's re-election campaign. Then, in Phases III and IV, the Nixon Administration removed most of the controls bottling up inflationary pressures. Costs and prices burst out and upward, as monetarists had predicted.

Monetarist theory - espoused most notably by University of Chicago economist Milton Friedman - states that the rate of change in the money supply determines both the rate of growth of the economy and the rate of price and cost increases.

From January, 1973, until Nixon resigned in August, 1974, consumer prices rose nearly 20%. The inflation was unprecedented for "peace-time" years in post World War II America. By the time Gerald Ford became President, prices were being inflated at a "double-digit" annual rate. The *Consumer Price Index,* the basic measure of price increases in goods and services used by Americans, rose more than 11% in 1974.

***

XXV. THE PROBLEM OF ECONOMIC GROWTH

Until recently, most Americans have believed that economic growth was inevitable in this land of progress. American parents have assumed that their children would enjoy an improved standard of living and greater opportunities than they had.

But early in its third century as an independent nation, the U.S. sees prophets of gloom and despair everywhere, sowing seeds of doubt. Some speak of the free enterprise system as stagnant. Others see a "no-growth" economy.

A study undertaken by the Department of Housing and Urban Affairs (HUD) concluded that "middle-class Americans have begun to feel that they have lost control over their destiny and even over their ability to maintain the standard of living they have obtained."

Pessimism about the future is not shared by some other researchers. Herman Kahn, director of the Hudson Institute, a Conservative think tank, sees mankind en route to the millennium.

In *The Next 200 Years,* Kahn foresees the greatest period of economic growth in history. At the same time, he predicts, population growth will decline and resources will go further. Looking ahead two centuries is extremely perilous. But Kahn predicts:[64]

"Two hundred years ago, almost everywhere human beings were comparatively poor and at the mercy of the forces of nature."

By 2176, Kahn says, world resources will more than adequately supply food, energy and raw materials needed by the global population, which he projects to be 15 billion, compared to some nearly six billion today.

Prospects for economic growth depend upon several factors. These include natural resources, the skill of the people who work in factories, mines, offices and fields. Existing capital goods are important, as is the structure of production. Perhaps most vital, are the nation's dominant values.

Population also is a key factor in economic growth. A nation may be too small, or more commonly, as is the case in Third World nations, too large for maximum development.

The Malthusian Law of Population

Thomas Malthus, clergyman and economist, wrote in 1798 that "...in a given state of the arts of production, population tends to outrun the means of subsistence."

Malthus observed that population increased geometrically while the food supply increased only arithmetically. Consequently, he saw disaster ahead.

Western technology made a mockery of this dismal prediction of scarcity. But Third World nations, often lacking skilled labor, improved methods of production and motivation to produce, stimulate "neo-Malthusians" to predict disaster for them.

Colonialism, for all its faults, introduced Western law and order and modern sanitation into Asia, Africa, Latin America and the Middle East. Some "positive checks" on population growth - famine, disease and high infant mortality - were eliminated. Today nations such as India and Pakistan concern themselves with reducing birth rates, the Malthusian "preventive check" on population growth.

In the U.S., gains in total production have meant economic growth and development. Although economists do not understand fully the processes of economic growth, they do see some reasons for American progress. These include the following:

(1) Technological development: The vast amounts of money invested in Research and Development (R&D) funds by American firms is a vital aspect of U.S. economic progress.

(2) Natural resources: The United States enjoys a moderate climate and enormous agricultural productivity. Although the energy crisis clearly indicated that no nation is entirely self-sufficient, the United States is abundantly blessed in natural resources.

(3) Skilled labor force: American workers traditionally have been well-trained and highly educated. With about 5% of the world's labor force, the U.S. produces about 25% of the world's supply of goods and services.

(4) Investment spending: Economic progress demands that oil, machinery, buildings and other capital equipment be replaced. We must be efficient to compete as a nation. U.S. industry, while suffering in certain areas, nonetheless has had adequate levels of investment spending. The nation's business firms, for example, have an average of more than $20,000 in plant and equipment investment in each production worker.

(5) Ideology: Although frequently attacked by its critics, American capitalism is an ideology which gives real incentives to produce. Both labor and capital have been motivated by the hope of material gain. Profits and higher wages have played key roles in powering this engine of economic growth.

***

XXVI. ECONOMIC STABILITY

Economic stability has been a major goal of U.S. policy since the New Deal. Although few quarrel with the objective, many debate the proper route to that particular destination.

Since the 1930s, government has been using two types of tools, called *automatic* and *discretionary stabilizers*, to minimize the impact of the business cycle on the economy.

The first of these requires no new political action. The second demands legislation or new policies by private enterprise. The goal is to avoid hyperinflation or depressions.

Perhaps most familiar of automatic stabilizers are individual and corporate income taxes. U.S. tax laws generally take more from those with growing incomes, less from those with shrinking incomes. But government spending tends to remain high even during recessions.

As a result, government tends to pay out more than it takes in from the public during a slump, limiting the impact of recession. Because U.S. tax rates are "progressive," taking a larger fraction of income as one moves into a higher bracket or taking a smaller percentage of income as one drops down the ladder, the impact of "boom or bust" is modified by this automatic stabilizer.

Social Security programs also have a *countercyclical effect*.

Government also issues more benefit checks as unemployment rises, despite the fact that employers, who finance the unemployment system, pay in less as workers are laid off. The opposite occurs when workers are rehired.

Government welfare payments also are larger during bad times than during periods of prosperity. These funds have limited the impact of "hard times" on millions of Americans.

What measures can government take to fight recession?

Congress willing, the usual prescription is a tax cut. Theoretically, additional after-tax income enables consumers to buy more. This generates new demand, leading to an upturn in business investment, jobs and consumer confidence in the economy.

In short, post-New Deal economics holds that a nation can "spend its way out" of a recession.

In addition to using taxes and public spending, the tools of fiscal policy, to fight recession, the United States also has monetary tools. The Federal Reserve Board (Fed) may seek to stimulate the economy by "easy money" policy or to dampen inflation by making it harder for consumers and businesses to borrow by squeezing the money supply - a "tight money" policy.

In the first case, credit is available to consumers on favorable terms. In the latter instance, the cost of money (interest rate) is increased to discourage borrowing.

Monetarist economists say, however, that this Keynesian-style monetary policy is counterproductive. Monetarists theorize that printing new money, an "easy money" policy, creates inflation by adding disproportionately to the supply of dollars in relation to the supplies of goods and services.

On the other hand, making the money supply grow slowly - "tight money," while it initially pushes up interest rates and may restrain borrowing, leads, the monetarists say, to less inflation and to better prospects for steady, long-term growth of the economy.

The private sector can play its part in achieving economic stability by planning investments to counter swings of the business cycle.

U.S. economic policy long has been characterized by Keynesian theory, although the nation has not balanced its budget over the business cycle, as Keynes advocated. In part, this problem stems from the *politics of economics.*

Since World War II, the American people have expected any administration to move energetically to fight recession.

Although some business leaders tend to view government intervention in the economy with misgivings - except when it benefits them - with defense contracts, import tariffs, quotas or direct subsidies, they have frequently co-operated with government in a common quest for a stable and growing economy.

Every post-World War II President, from Truman to Clinton, has had to promote the well-being of American business. They also have tried to promote a competitive economy while preventing abuses of economic power.

***

XXVII. MONETARY POLICY

One of the least understood governmental institutions in economic policy is the Federal Reserve Board. It is, however, a body that affects the economic life of all Americans by setting and carrying out monetary policy.

Commonly known as the *Fed,* the board seeks to regulate U.S. money supply. There are a number of ways of measuring how much money is in the U.S. economy but the two basic measures which the Fed reports each week, are M1 and M2.

M1 estimates the value of all currency - Federal Reserve notes (dollar bills) and coins - in circulation together with the value of checking accounts in the banks.

M2 is M1 plus the value of savings accounts and certificates of deposit of less than $100,000.

The amount of money in circulation depends primarily on the wishes of individual businesses and consumers about how much cash they want on hand, rather than in banks.

Growth or shrinkage of checking accounts is determined largely by Fed policy. Private banks create checking accounts when making loans and investments. When, for example, a customer receives a $25,000 loan, his bank establishes an account for that amount. He then can transfer those funds to other banks which can, in turn, make loans and create further deposits.

Banks also can invest idle funds in government securities. They create additional deposits in other banks when they pay for these securities. Of primary importance is the ability of banks to lend out more money than they are holding in cash. As long as the inflow and outflow of funds roughly matches, banks need not keep all of their funds on hand to cover withdrawals.

The Fed requires its member banks to keep only a part of their deposits as reserves. The system thus is called a *"fractional reserve"* system.

This process may seem quite complicated, but it really is quite simple. It does not need a central banking authority to keep order. The decision to create the Federal Reserve System was in fact triggered by the Panic of 1907.

In that year, the U.S. saw an end to a decade of unrivaled prosperity. Several railroads went into receivership. Some 13 New York City banks failed and unemployment and wage cuts occurred. It was largely a bankers' panic brought on by speculation and rash management. Conservatives, however, said that the panic was due to the business community's lack of confidence in the policies of President Theodore Roosevelt.

Roosevelt denied this and tried to avert future bankers' panics by signing the *Aldrich-Vreeland Act*. This 1908 law provided that the U.S. Treasury Department might lend hundreds of millions of dollars to banks hard hit by monetary inflexibility.

Furthermore, a commission, headed by Senator Nelson W. Aldrich (R., N.Y.), was established to find some solutions to the nation's banking and currency problems. This was the forerunner of the Fed, initiated in 1913.

The Fed sets the percentages of reserves that member banks must keep on deposit to cover future withdrawals. It can, therefore, either increase or decrease the number of loans and investments banks can make.

Federal Reserve banks buy and sell Treasury securities, releasing or taking in huge sums of cash in the process. At times, the Fed lends money to member banks against their government securities or other assets at a *"discount rate"* (interest rate) set either to encourage or discourage borrowing and lending.

To encourage independence of the Fed, Congress empowered the President to appoint members of its board of governors to 14-year terms.

As a practical matter, the Fed chairman meets frequently with the President. Because few board members want to serve 14 years, vacancies occur fairly frequently.

Fed policies affect the interest rates consumers pay for car loans, mortgages and business loans.

Presidents often criticize Fed monetary policy, but the seven-member board has pursued a relatively independent course. Its power comes from Congress, not from the President.

Whether an independent body should have power to decide such vital economic questions without concern for public opinion is debatable.

During his 1976 presidential campaign, Jimmy Carter said:

"While the Federal Reserve Board should maintain its independence from the executive branch, it is important that throughout a President's term he have a chairman of the Federal Reserve whose economic views are compatible with his own. To insure greater compatibility between the President and the Federal Reserve chairman, I propose that, subject to Senate confirmation, the President be given the power to appoint his own chairman of the Federal Reserve who would serve a term coterminous with the President's."

Some criticized Carter's proposal because it seemed to ignore the original rationale for the independence of the system. They categorically rejected the notion that a President should have his own Fed chairman who would serve, like any other presidential appointee, at the pleasure of the Chief Executive.

Carter's proposal was directed specifically at former Fed chairman Arthur F. Burns. Burns, a Nixon appointee, long considered inflation to be the greatest economic threat to the United States. Burns had said:

"I do not believe I exaggerate when I say that the ultimate consequence of inflation could well be a significant decline of economic and political freedom of the American people."

Inflation, Burns argued, was the worst social evil because it threatened the very existence of American capitalism and freedom.

But Burns, in fact, led the Fed to expand the money supply rapidly in the early 1970s. Perhaps he sought to help the Nixon Administration bring the nation out of the stagnation of 1970-71. Perhaps he sought to aid Nixon in his bid for re-election in 1972.

The economy did expand rapidly in 1972, but the ballooning of the money supply also did much to increase the rate of inflation and helped to bring on the severe slump of 1974-75.

President Carter decided to replace Burns as Fed chairman with G. William Miller, head of the Textron conglomerate. Published reports indicated that White House economic adviser Charles Schultze believed Burns' policies were putting the brakes on the economy and preventing creation of new jobs.

But Miller said that if Carter wanted new programs, "...we must carve the money out of existing ones. National health insurance must be deferred. It is quite inflationary at this point. We cannot let the deficit creep up."

One of the most powerful men in Washington at the end of the 20th Century – some say second in influence behind only the President – is Alan Greenspan, chairman of the "Fed." Under Greenspan, described by some other economists as a subtle pragmatist, the American economy has prospered.

Appointed originally by President Reagan, Dr. Greenspan also has served in that role under Presidents Bush and Clinton. Economic growth in recent years has been about 4%, unemployment only slightly above that, and the inflation rate a very modest 2%.

***

XXVIII. FISCAL POLICY

Fiscal policy is one of the basic tools used by government to stabilize the U.S. economy. This involves the power to tax and spend and to use the budget either to speed up or to slow down the economy.

Historically, the use of monetary policy (control of money supply and interest rates) to counter inflationary booms and recessionary busts is not new. But the idea that the Federal budget, taxes and spending levels should be used as tools to stimulate or calm the economy is primarily a post-New Deal development.

Consumers, rather than government, account for most of the nation's spending. Increased consumer spending can pull up prices, particularly if materials are scarce. One way to deal with "excess" consumer demand is to increase taxes and soak up consumer purchasing power.

Conversely, in periods of recession or depression, government can reduce the tax burden on consumers and business. Workers, with less money taken from their paychecks, probably will spend more. This increased consumer spending may help to combat a slumping economy.

When government takes less taxes from business, funds can be freed for investment, hiring of additional workers and increased production, all of which stimulate a sagging economy.

As demand shrinks, supplies are in surplus. Unable to sell their goods at the profit expected, firms may cut production to reduce the surplus. Workers will be laid off. With uncertain economic futures, they cut spending to a minimum. This can lead, if unchecked, to recession or depression.

The auto industry, for example, was thrown into chaos by the sharp, steep and sudden increase in the price of imported oil in 1973-74. Many Arab oil exporters, some of the most important members of the Organization of Petroleum Exporting Countries (OPEC), further compounded the problem with an embargo against export of their oil to the United States.

Unsure of continued oil supplies, and faced with possible gasoline rationing, American consumers quit buying Detroit's "gas guzzlers."

With cars stacking up in dealer showrooms and in storage lots, auto manufacturers cut production, laid off workers and moved slowly toward production of smaller cars. Once the crisis appeared to be over, U.S. consumers began to buy again and the 1977 model year was a big success.

In recent years, some economists have urged tax cuts during periods of prosperity to maintain economic health at a high level. The legacy of the Kennedy-Johnson "New Economics" reflects this line of thought.

Some Conservatives argue that such tax cuts ultimately bring inflation. Some economists question the usefulness of fiscal policy; most recognize its limits.

Economist Elbert V. Bowden comments:[65]

"Fiscal policy is not a 'fine tuning knob' for economic stabilization. When we try to solve our ...problems with fiscal policy, we go after them, not with a surgeon's scalpel, but with a meat cleaver. Any time 'meat-cleaver action' is justified, fiscal policy will work. It may work too much or too little, but it will work."

Government since the 1930s has attempted to use fiscal policy to promote economic stability, full employment and stable prices. It has relied heavily on budgets and taxing and spending powers.

But it also has tried to create, particularly during periods of high unemployment, public works jobs for those who cannot find employment in private enterprise. Public jobs created by Franklin D. Roosevelt's administration set a precedent for those created during the Carter Administration.

President Carter, concerned by continuing high levels of unemployment (6.1% in February, 1978) and the possible drag of already enacted increases in Social Security taxes, proposed a $25 billion tax cut in January, 1978.

Carter, however, tied his tax cut to tax reforms, which encouraged strong resistance in Congress and in the country. House Speaker Thomas P. (Tip) O'Neill Jr. predicted:

"If he winds up with 50% of reform, he's done well."

House Republican leader John Rhodes of Arizona argued that the $25 billion cut was not enough and that it should have been $51 billion "to accomplish things necessary for the economy."

Some economists are concerned that government at times appears to pursue inconsistent economic policies. Congress and the President may plan stimulative measures. They may wish to cut taxes and increase spending and create more federally funded jobs.

But the chairman of the Federal Reserve Board then may warn that constant price rises force the Fed to keep the money supply tight, thereby retarding construction and hiring.

***

XXIX. THE FEDERAL BUDGET

"To most people - college students included - the national budget is B-O-R-I-N-G. To national politicians, it is an exciting script for high drama. The numbers, categories and percentages that numb normal minds cause politicians' nostrils to flare and their hearts to pound. The budget is a battlefield on which politicians wage war over the programs they support."
--Political scientist Kenneth Janda, *The Challenge of Democracy*.

The Federal budget is more than a vast collection of charts and graphs, facts and figures, although it assuredly is that. President Clinton's printed budget for 1999 was more than 1500 pages long. It is pre-eminently a statement of national values and priorities, determining who gets what. Consequently, it is the result of a complex process of pressure politics and compromises on Capitol Hill.

The President sends his budget to "The Hill" in January after the Office of Management and Budget (OMB) has done an enormous amount of work, screening proposals from departments, bureaus and agencies to insure that they are consistent with the President's policies.

It applies to the next *fiscal year* (FY), the accounting period the government uses. Currently the fiscal year runs from October 1 to September 30. Since the budget carries the name of the year in which it ends, FY 2000 applies to the 12 months from October 1, 1999 to September 1, 2000.

Before 1974, Congress would enter into exhaustive debate in the House and Senate. A series of bills enacted on a piecemeal basis would authorize and fund programs.

Finally Congress recognized that its budget-making process was haphazard. Frequently it did not consider questions of overall revenues or spending while examining legislation. Congress passed the *Budget Control Act* of 1974.

The law required Congress to establish overall spending limits while looking at the budget as a whole and setting its own spending priorities.

Government also uses the budget as an important tool of fiscal policy, either to speed up or slow down the economy. Stimulation is the tool if the economy is in a recession. Restraint is used if inflation is the paramount concern.

Some Presidents, notably Reagan and Bush, preferred to delegate economic policymaking to subordinates in their administrations. Clinton appears to have been more involved in the budgetary process. Some Presidents, notably Nixon and Bush, were interested primarily in foreign policy. Others, including Clinton, appear to be more involved in economic policy.

In January, 1978, Jimmy Carter sent Congress a budget of $500.2 billion for fiscal 1979, an accounting period which began on October 1, 1978 and ended September 30, 1979.[66]

Governmental functions then and now are divided into a number of categories such as national defense, health, education and welfare and energy. Carter's proposals called for higher federal spending levels for defense, energy and education, but no larger outlays on entirely new programs.

It was difficult at the time for most Americans to comprehend $1 trillion - particularly with its 12 zeroes. Carter's budget of $500,174,000,000 was only slightly more than half of that amount. But House Speaker Thomas P. (Tip) O'Neill (D., Mass.) told reporters, "I would like to see us under $500 billion."

Veteran members recalled President Johnson's efforts some 14 years earlier (1964) to keep the budget under $100 billion. Rep. George D. Mahaon (D., Texas), chairman of the House Appropriation Committee, recalled that a $6 billion budget was considered exceptionally high when he entered Congress in 1935.

The numbers in Carter's budget no longer shock anyone, given the enormous inflation that occurred during his presidency and the constant growth of Federal spending in the past two decades. Bill Clinton's budget request for FY 1997 was for more than $1.6 trillion.

By May, both House and Senate approve a first budget resolution, setting overall revenue and spending targets. This is complicated, as it was during the Carter years, by the President's tax cut and tax reform proposals. Carter's budget estimated $439.6 billion in receipts against the $500.2 billion in outlays, leaving the U.S. government in the red by some $60.6 billion.

By September, the Congress approves a second budget resolution, setting more binding limits on U.S. spending.

In the past generation considerable debate has occurred on the continuing large deficits. The late Sen. Edmund Muskie (D., Maine), chairman of the Senate Budget Committee, warned his colleagues that:

"Liberals must start raising hell about a government so big, so complex, so expensive and so unresponsive that it's dragging down every good program we've worked for...Why can't Liberals start hacking away at the regulatory bureaucracy where it keeps costs up and competition away? Why can't Liberals talk about fiscal responsibility and productivity without feeling uncomfortable?"

The prospect of a $60.6 billion Federal deficit on top of a $61.8 billion deficit caused many economists and others to question whether Carter's promise to balance the budget for the fiscal year beginning October 1, 1980, was possible.

The Federal debt was an estimated $752 billion in July, 1977. Even in a $2 trillion economy this caused many to pause. The overwhelming majority of the American people, according to a Gallup Poll of April, 1976, believed in a balanced budget.

The Federal debt does not alarm everyone, however. Some economists look at it in relation to *Gross Domestic Product (GDP)* - the total market value of all final goods and services produced annually within U.S. boundaries, whether by American or foreign-supplied resources. In 1945 national debt was about equal to GDP, then about $200 billion. By 1960, Federal debt had increased to $250 billion, but GNP was $513 billion. Due largely to the enormous deficits of the Reagan years, and continued deficits during the Bush and Clinton presidencies, the national debt stood at about $5 trillion in 1998. GDP was then about $8.5 trillion.

In 1998, however, the Federal budget reported a surplus of $69 billion, the first surplus since 1969, and reduced the Federal debt by more than $50 billion.

The U.S. moved from deficit to surplus because the Republican Congress insisted in the late 1990s – and the Democratic president agreed – that spending growth has to be restrained.

It is always perilous, particularly as a nation nears the elections of 2000, to take political statements as facts. Nontheless, President Clinton was predicting in late June, 1999, that by the year 2015, the U.S. would owe no money to investors.

Economic Conservatives often argue that government, like individuals, can abuse its borrowing privilege. But the results of debt abused by individuals and by government are very different. Some economists believe that the primary burden of the public debt is the annual interest charge accruing as a result. Interest payments on the national debt constitutes the fourth largest item in the Federal budget, trailing behind only Social Security, National Defense and Public Health programs.

Federal government has a much greater opportunity to be irresponsible in its spending than does the individual. Congressmen and Presidents can make political capital by following the slogan: "Tax and spend, elect and elect," reportedly first uttered by Harry Hopkins, one of Franklin D. Roosevelt's key White House political advisers.

***

XXX. INFLATION: FROM CARTER'S NIGHTMARE TO "NO WORRY"

Inflation was the most troublesome economic problem facing the U.S. during the decade of the 1970s. During this period, price levels increased rapidly.

When Gerald R. Ford assumed the presidency in August, 1974, the annual rate of inflation in the United States was at the "double digit" level of more than 10%. Although it slowed for a time thereafter, the cost of living remained the principal economic concern of the American public, according to George Gallup and other pollsters.

W. Michael Blumenthal, Secretary of the Treasury under Jimmy Carter, warned that inflation could top the 6.8% level of 1977 unless business and labor restrained themselves in seeking price and pay increases.

In launching an anti-inflation drive, President Carter said, "Inflation has become embedded in the very tissue of our economy."

While high inflation rates are unacceptable to Americans, the problem is global, rather than merely national. United States' increases in cost of living are significantly less than those of other nations. During the mid-1970s, for example, Japan and many European nations experienced inflation rates of more than 20%. Prices also rose at annual levels of from 20% to 50% in some of the less developed countries (LDCs).

Historically, inflation has occurred during wartime when government spending escalates and private spending remains at high levels unless major tax increases "soak up" consumer purchasing power. Congress did not enact major tax increases during either the Korean or Vietnamese wars.

Although past inflations have been relatively brief, recent inflation seems to be chronic. It has not been followed, economists note, by gradual price declines as in earlier years.

What was the cause of U.S. inflation and hyperinflation during the Carter era?

Economists and others do not agree on the basic cause of the problem, but do offer a variety of perspectives. War, as noted, is usually a major contributing factor. So was the oil embargo which followed the 1973 Arab-Israeli war.

Massive increases in energy prices seriously disrupted U.S. and other economies. Between October 1, 1973, and January 1, 1974, world oil prices climbed from $2.50 per barrel to more than $11 a barrel. Again in 1979-80, prices climbed rapidly, this time to $35 per barrel.

William Simon, then Treasury secretary, told a congressional committee that half of the rise in the wholesale price index for the year ending in mid-1974 was due to increased oil prices alone.

Food shortages, caused by bad weather and reduced crops, caused a sharp increase in basic agricultural prices.

Another contributing factor was demand for wages that outstripped gains in productivity. While it is understandable that workers wanted more money to pay for increased cost of food, fuel, housing, clothing, autos, medical care and many other things, such increases were passed along to consumers in the form of higher prices. Business had to meet its increased labor costs. It appears to have been a vicious cycle.

Some Conservatives consider high levels of Federal spending to be a major cause of inflation. Modern governments play a more active role in producing demand for goods and services than they did long ago.

Most economists concede that U.S. policies of spending more money than it garners in tax revenues is highly inflationary.

In a speech to the American Society of Newspaper Editors (ASNE), President Carter promised that government would make its own contribution to anti-inflation policy by holding the Federal deficit to the $61 billion he had budgeted.

Carter also announced that he would limit pay raises for Federal employees and threatened to veto legislation which might push the United States deeper into red ink.

The President also promised to make Federal regulation of business less burdensome. Vast amounts of required bureaucratic paperwork frequently are inflationary, business economists believe.

But can government control inflation? A number of Presidents have tried to attack the problem with a variety of techniques.

Kennedy tried to fight inflation by "*wage-price guidelines*" intended to measure gains in productivity and to pressure unions and management to relate wage and price increases to productivity indices. These guidelines failed.

Johnson tried *"jawboning"* as well as guidelines. This technique was designed to mobilize public opinion against "excessive" increases in wages and prices.

Nixon imposed a program of direct controls over wages and prices in August, 1971. But the controls were abandoned shortly after return of prosperity in 1972.

Ford launched a highly publicized Whip Inflation Now (WIN) program. It proved short-lived. President Carter, rejecting direct economic controls over wages and prices, said:

"I can't imagine any circumstances under which I would favor wage and price controls other than a national emergency like all-out war."

Union leaders and some political cartoonists suggested that Carter sounded much like his predecessor in his speech to the American Society of Newspaper Editors. During the 1976 campaign, however, Carter had listed inflation as a less serious threat to the nation's economy than unemployment.

History, however, makes it clear that inflation was one of the most important factors in Carter's failure to win re-election in 1980. It was, according to political historian-journalist Theodore White, one of the two factors most responsible for Reagan's victory over Carter. The other was the national humiliation experienced by the United States in Iran when Ayatollah Ruhollah Khomeini's regime held 52 American diplomats and Marines hostage for 444 days.[67]

The Reagan-Bush years, while characterized by massive increases in the national debt, were years in which inflation was not a matter of much concern. Mild inflation is seldom a matter of concern, and may, in fact, be a political plus. Inflation thus far in the Clinton presidency also has been mild. What about a renewed threat of inflation in the last few years of the 20th Century?

Inflation during Clinton's first two years in the White House was well under control, with rates of 3% in 1993 and only 2.6% in 1994. The Open Market Committee of the Federal Reserve Board was nonetheless inclined to take a "wait and see attitude." There were some hopeful signs for the U.S. economy. Among them:

(1) The unemployment rate for 1993 was 6.8%. It dropped to 6.1% in 1994. On the eve of the presidential election of 1996, it was running at less than 6%. Construction starts on homes and apartments jumped to their highest levels since February, 1990. U.S. plants manufactured 10.7 million cars and trucks in 1993, a 9% increase from the 1992 model year.

(2) Mortgage rates at the end of 1993 were at near 25-year lows, stimulating sales of homes and new home construction. This potentially means growth in sales of appliances, furniture and other home-related goods and services.

(3) Inflation, some argue, has been held down primarily by collapsing crude oil prices and a 23% drop in tobacco prices. In the 12 months between December, 1992, and December, 1993, consumer prices rose only 2.7%, wholesale prices only 0.3%. In 1992, consumer prices rose only 2.9%. Inflation did not appear to be a particularly bothersome issue entering the presidential election campaign of 1996.

This underscores a point: economists experience different *"worry cycles,"* being concerned about different economic issues at different times. This also is true of society as a whole and politicians, always concerned with the "pulse of the public" (as it effects their re-election prospects) in particular. Inflation was the most important issue of the 1970s. Americans at the end of the 20th Century view other issues as more important.

***

XXXI. UNEMPLOYMENT

Although few Americans in 1999 consider inflation the nation's principal economic problem, many rate unemployment at the top of their "worry list."

What is *unemployment?* Although it would seem easy to define, it is not. Some would consider anyone without a job as unemployed, but the government does not. Some unemployed persons, such as students, those more than 65 years old and "housewives," are voluntary "unemployed."

Technically an "unemployed" person is not merely someone without a job. It is someone who is actively seek employment, but unable to find it.

There are, economists say, three major types of unemployment: frictional, structural and cyclical.

Frictional unemployment is short run in nature and is "normal." It occurs when workers voluntarily leave one job for a better one. It is not considered a major economic problem since it is transitional.

Structural unemployment is long run in nature and is a serious problem. It results when demand for a particular kind of labor is low in relation to its supply in particular sectors of the economy. This kind of unemployment may be due to a lack of skills or because those seeking jobs do not have the kind of abilities that employers want.

Government has attempted to retrain some workers made jobless by the introduction of new kinds of technology. But job retraining is costly and the question always arises: For what jobs shall the unemployed be trained? Care must be taken to avoid training workers for jobs which will not long exist after workers are ready for them.

Cyclical unemployment is caused by fluctuations in the business cycle. It results from decline in demand for goods and services. Whenever the unemployment rate exceeds 4%, most economists believe that cyclical unemployment is to blame.

Each month the Federal government announces its employment and unemployment statistics, but these figures must be viewed with caution. Economists and political leaders view not only the number and percentage of unemployed, but also the number of employed, now at an all-time high.

Each month headlines report an unemployment rate. But the number of Americans who have jobs is also vitally important. Is the economy creating new jobs? If so, what kind are they?

Unemployment, of course, does not affect all groups equally. Black unemployment for years has been about double that of white unemployment. Teenagers traditionally have high rates of unemployment. Lacking experience and/or skills, they often are marginal workers, among the last to be hired and the first to be fired.

Unemployment also hits different groups of workers in different ways. Blue collar and service workers are twice as likely to be unemployed as white collar or farm workers.

What can be done to cope with the economic and social effects of unemployment? The economic consequences include a decline in production of goods and services. Clearly, idle human resources represent an economic waste.

Perhaps, however, the greatest misfortune of unemployment is its social effects. It threatens the social fabric, the family, human relations, and the worker's own self-respect and confidence.

Since the 1930s and the New Deal of Franklin D. Roosevelt, the public has looked to Washington for aid in dealing with joblessness. The Congress, in 1946, passed the Employment Act, establishing full employment as a national goal. In 1978, Congress enacted the *Humphrey-Hawkins Full Employment and Balanced Growth Act*. Humphrey-Hawkins set an unemployment target of 4% or less by 1983. Although endorsed by union labor and President Carter, the law was questioned by many others, including the President's own Council of Economic Advisers.

The Council said that "...it is unlikely that a 4% unemployment rate can be achieved through aggregate demand policies alone without at the same time causing a significant increase in the rate of inflation."

Economist Herbert Stein wrote in his book, *Presidential Economics*:[68]

"Every informed person knew that the whole idea - including the 4% unemployment - was nonsense and that the bill could be stomached only on the assumption, which proved to be correct, that it would be forgotten as soon as enacted. But no political person, certainly not one dependent on the traditional Democratic constituency, found the courage to say so. Although the Carter

Administration tried to moderate the bill in the end, it also had to act as if the 4% goal was in some long-run sense reasonable."

Some economists believe that there is an inevitable "trade-off" between unemployment and inflation. In 1958, Professor A.W. Phillips of Australian National University developed the so-called *Phillips Curve*, a major factor in employment policy.

In essence, Phillips' doctrine holds that if you want less unemployment, you must accept more price increases. Conversely, if you want less price increases, you must accept more unemployment. Conventional wisdom held that high rates of inflation inevitably are accompanied by low rates of unemployment and vice versa.

Congressmen and Presidents have repeatedly said that we must accept high unemployment rates in the future if inflation is to be brought under control.

Not all economists accept Professor Phillips' curve. Arthur Burns, former chairman of the Federal Reserve Board, for example, said:

"Whatever may have been true in the past, there is no longer a meaningful tradeoff between unemployment and inflation...I believe that the ultimate objective of labor market policies should be to eliminate all involuntary unemployment. This is not a radical or impractical goal."

It remains for the President and the Congress to formulate national job goals. Part of the recent problem has been demographic, a massive number of teenagers entering the job market. With lower birth rates becoming a factor in the future, unemployment may be a less serious problem in the future.

***

XXXII. POVERTY AND WELFARE REFORM

"Poverty is one of those emotionally-charged words that can trap you if you are not careful. Much needless soul-searching can be avoided if it is recognized at the outset that there is no objective definition of poverty any more than there is an objective definition of art or beauty. The standards of poverty are established by society...they vary from place to place and they differ from time to time." So said Herman Miller of the U.S. Census Bureau in trying to point out the difficulty of defining poverty.

Who really is poor? How can one say that one family is poor and that another is not? Poverty is a relative and subjective term and depends upon a number of variables. Some suggests that individuals and families are poor when they are unable to meet their *basic needs*.

Family size has much to do with poverty. So does age, place of residence (urban or rural area), opportunity to grow one's own food and consumer price increases.

History makes a difference too. An urban family of four making $6000 in 1978 was considered poor by government standards. It would have been considered well off in 1962 when the *"poverty line"* was $3000. In 1999, the official government definition of a "poor" family (urban family of four) was $16,700. For single individuals it was $8240.[69]

Poverty also means something very different in the U.S. than it does elsewhere in the world. By American standards, 75% of all the families in Great Britain could be considered poor. Millions of people living in less developed countries (LDCs) in Asia and Africa would be ecstatic if they could raise their standards of living to match that of the American poor.

In the poorest county in Mississippi, our poorest state, a majority of the poor own television sets. Nearly a majority (46%) own automobiles and almost 40% own washing machines. Most Americans tend to define poverty in terms of dollars. So does the U.S. government, as indicated above.

Poverty also has been mixed with the civil rights movement because of its prevalence among blacks. Blacks often are unable to get good educations and are victims of family disorganization and discrimination.

In 1993, it was reported that about 12% of the nation's white population was poor, compared with 33% for Blacks and 31% for Hispanics. Overall, some economists estimate the present national population of poor people at about 10-15%.[70]

In 1993, more than one-third of households headed by females were poor. Education is of central importance. Only about 2% of college-headed families are poor while those with eighth-grade educations or less are three times as likely to be poor as families headed by high school graduates.

There is much myth about poverty and welfare in the U.S. Despite a widely-held belief to the contrary, most poor people are not on welfare. They work - but at poorly paying jobs. Many of them are migrant workers who are able to find only part-time employment or simply lack job skills.

Some sociologists and others have written that poverty itself is the major cause of poverty. Although that sounds very much like a circular argument, many agree that there is a *"culture of poverty,"* an environment of social despair which tends to perpetuate itself.

Some children, for example, grow up in the demoralizing atmosphere of the city slum. They quickly lose hope as their parents did earlier. They often lack adequate nutrition and health care. They often cannot compete with middle class children in school who have books in their homes, become discouraged and drop out in high numbers.

What, if anything, should government do to improve the conditions of the poor? Comparatively few Americans question the need for public welfare programs for the poor, although there is enormous debate over their form and extent.

Liberals and Conservatives disagree about what should be done to minimize or eliminate poverty in America. Liberals argue that poverty is such a complex problem that only government can cope with it.

Some Conservatives believe that government "interference" has slowed economic expansion of the free enterprise system. They believe that the economy could grow faster and create more jobs if government interfered less with business.

Some economists, such as Heilbroner, argue that the United States has the capacity, but not the will, to eliminate poverty. Heilbroner argues that a shift in government spending of about 1% of the Gross National Product (GNP) could reduce poverty to a minimum.

Liberals and Conservatives agree that existing government poverty and welfare programs do not work as well as they might. President Carter recognized this in urging a welfare reform program. Among other things, the President wanted to move toward a guaranteed family income through a *negative income tax (NIT)* and to replace food stamps with cash payments.

The present Federal tax system calls on families to "subsidize the government" when their incomes rise above a certain point. NIT would call on government to subsidize families whose incomes fall below a certain level, to be determined by Congress.

Carter's proposal ran into trouble on Capitol Hill. Al Ullman (D., Ore.), powerful chairman of the House Ways and Means Committee, believed that the President's proposals would accomplish little genuine reform while drastically accelerating inflation.

Welfare Reform

The poverty-welfare reform issue is back on the political agenda in 1996. During the 1992 campaign, Clinton and Al Gore, his vice presidential running mate, published a little book.[71] In this campaign document, they pledge to "end welfare as we know it." Main points:

(1) Empower people with education, training and child care they need for up to two years, so that they break out of the cycle of dependency; expand programs to help people learn to read and get their high school diplomas; and ensure that their children are cared for while they learn.

(2) After two years, require those who can work to go to work, either in the private sector or in community service; provide placement assistance to help everyone find a job; and give people who can't find employment a dignified and meaningful community service job.

(3) Actively promote state models that work.

Liberals as well as Conservatives often refer to the present system as a "mess."

Under Federalism, states often have served as laboratories where social experiments have been tried. Several states, such as Michigan and Wisconsin, in recent years have launched "welfare to work" (workfare) plans, designed to encourage welfare clients to move from public assistance to jobs. They have been given child care and transportation subsidies in return for entering training or job programs.

Clinton has proposed to make work more attractive by expanding the *Earned Income Tax Credit,* effectively increasing wage rates. Government would make up any difference between a family's earnings and the poverty level. He also would force fathers who abandon their children to support them. This could be done by using the machinery of the Internal Revenue Service to help collect child support payments.

Clinton also proposed a "national deadbeat" databank to enable law enforcement officers to track down negligent parents more easily. He pledged to work for legislation to make it a felony to cross state lines to avoid paying child support. Deadbeat parents would be reported to credit agencies, so that they can't borrow money for themselves when they're not taking care of their children.

During his 1996 campaign, Clinton hailed Wisconsin's *mandatory work plan* and promised to support it.

The AFDC *(Aid To Families With Dependent Children)* program is administered by the states, but financed in part with Federal grants. Aid is made available to those children who do not have support of a parent, usually the father, because of divorce, disability, death or desertion.

Today AFDC remains one of the most controversial of welfare programs. Critics argue that it encourages family dissolution, promotes illegitimate births, supports a "culture of poverty," is ridden with fraud and is destructive of the work ethic.

In 1988, the Democratic Congress passed and Republican President Reagan signed the *Welfare Reform Act*. It requires states to set up Job Opportunities and Basic Skills programs through which AFDC parents will be offered basic and remedial education, literacy classes, jobs skills training, job readiness activities and job placement.

States also are required to provide transitional child care and Medicaid coverage for a year to families switching from welfare rolls to payrolls. The thinking behind this provision is that it would lessen incentives to remain on welfare.

All states also are required to offer welfare benefits to qualified *two-parent families* when the main wage earner is unemployed. The law is aimed at prevention of family break-up associated with AFDC.

Both President Reagan and congressional leaders hoped that the law would help to end the *"culture of welfare"* in which dropping out of school, having children and going on public assistance have become a "normal" way of life for a segment of the population.

The 1988 act apparently had not achieved its objectives by 1996. Bob Dole and Bill Clinton both again were promising to "change the system." The Republican-controlled 104th Congress had passed reform legislation. The bills prohibited welfare to minor mothers and denied increased AFDC payments to anyone who had additional children while on welfare. AFDC recipients were required to work. Clinton used his veto. The matter once again was before the voters.

Facing a re-election bid, Clinton confronted the political necessity of compromising with the Republican-controlled Congress. When it passed Welfare Reform legislation for the third time on the eve of the election, the President signed the measure into law.

Basically the bill gave more control over welfare to state governments. States determine, for example, who will receive what kind of benefits. Aid-to-Families With Dependent Children (AFDC), one of the least popular welfare programs with the public, was abolished. It was replaced by a new program, providing Federal funds to states (block grants) to meet cost of welfare programs. Although the U.S. maintained control over food stamp programs, benefits were reduced.

The basic purpose of Welfare Reform was to reduce spending by all governments by imposing work requirements and curtailing payments to unmarried teenage mothers. One critic of the new law, Sen. Daniel Patrick Moynihan (D., N.Y.), called it not welfare reform, but welfare repeal.

Democratic party failure to regain control of Congress in 1996 and 1998 meant no "undoing" of the reforms.

***

XXXIII. SOCIAL SECURITY

Social Security emerged during the middle 1970s as a major political issue for the first time since it was passed in 1935 and debated during the presidential campaign of 1936. In the last few years of the 20th Century, we are debating it again.

When Congress first enacted the legislation, some predicted Social Security would bankrupt the U.S. Treasury and the moral fibre of American citizens. The Republican Party platform of 1936 and GOP presidential nominee Alf Landon of Kansas called for repeal of the act. They called it "socialistic."

The U.S. was one of the last major industrial nations to enact Social Security legislation. Germany, under Chancellor Otto von Bismarck, was the first country to introduce a workers pension program in 1889. Bismarck clearly was no socialist. He was, however, a first rate political strategist and wanted to undercut the appeal of the German Social Democratic Party. In 1911, Great Britain established a similar retirement program and by the time of the First World War, Social Security systems were in effect in all major European countries.

After Franklin D. Roosevelt swept 46 of 48 states in his 1936 re-election campaign, Social Security was no longer much debated. Republicans, as well as Democrats, embraced it. Wendell Wilkie (1940), Thomas E. Dewey (1944 and 1948) and Dwight D. Eisenhower (1952 and 1956) all pledged not only to continue the program but also to improve its administration. In short, Social Security became the bedrock of the American welfare state, supported by both major political parties.

Over the years the American public has regarded Social Security as a pillar of the American way of life. Benefits were improved substantially and new programs were added. But to pay for these improvements, taxes on both workers and employers were increased substantially.

When Social Security legislation first was enacted, the payroll tax on workers was 1% on the first $3000 earned. This was matched by a 1% payroll tax on the employer. The maximum contribution by a worker was only $30 per year. Workers who reached the age of 65 beginning January 1, 1942, were given minimal benefits.

Single workers could receive $22 per month and married workers $36. Those workers who received very low benefits from Social Security might qualify for Old Age Assistance, a public assistance program for the elderly poor included in the Social Security Act.

Shortly after enactment of the law, a citizen's advisory council was established to review the Act periodically. Social Security grew and became enormously popular with the American public over the years. After the Landon disaster of 1936, no politician dared to challenge it. The old saying in Congress is "Social Security is like the third rail in the subway. Touch it and you die."

Over the years, more groups became eligible for Social Security benefits and they soon eclipsed Old Age Assistance. The law was amended to correct gender problems by changing provisions that treated men and women differently. Disability insurance was incorporated in 1956 by the Eisenhower Administration. By the time of the Kennedy-Nixon debates in 1960, some 60% of American workers were covered by Social Security. By 1970, some 82% were covered; by 1980, some 90%.

Medicare and Medicaid

In 1965, as Part of Lyndon Johnson's Great Society, Congress enacted Medicare, a program of health care for Americans over the age of 65. Not just poor persons over 65 who needed help paying their medical bills, but ALL Americans over 65 were covered. By 1980, some 26 million Americans were receiving Medicare, which Liberals had hoped would presage a system of universal national health care insurance.

By 1993, the number had grown to 39 million insured for hospitalization under Medicare (Part A) and more than 34 million insured for physicians care (Part B). Despite the fact that hospital benefits were limited and deductible levels were substantial under Part B, medicare became popular with the elderly. For those who were very ill, elderly and in need of help in paying for long-term nursing home care, Medicaid, another 1965 Great Society health program, came into play. To become eligible for medicaid, however, one had to be genuinely poor, and this often involved selling one's home and exhausting other assets until he or she could qualify for nursing home services.

Medicare was to be financed by a portion of the same withholding tax that funded Social Security.

Indexing of Social Security Benefits

A major change occurred in Social Security law in 1972, when Wilbur Mills, then chairman of the tax-writing House Ways and Means Committee, decided to seek the Democratic Party's presidential nomination. Mills needed an issue and thought that he had found one in "indexing." Only a year earlier Mills had opposed automatic increases in Social Security benefits if the cost of living rose beyond a certain point.

When, however, Mills learned that New Hampshire, the first state to hold a presidential primary, had the second highest percentage of senior citizens in the nation, he became a champion of indexing. He announced that he now favored increased Social Security payments, related to inflationary rises in the Consumer Price Index (CPI). Congress passed indexing that summer, although Mills' campaign already had collapsed.[72]

But increased benefits also meant increased taxes. Under the massive increase in taxes voted by Congress in December, 1977, the payroll tax went to 6.05% on the first $17,700 for both employer and employee and rates rose sharply over the years. In 1994, the rate rose to a flat 6.2%. Salary subject to tax increased from $22,900 in 1979 to $31,800 in 1982.

The "cap" rose over the years until the first $60,600 of one's salary is now subject to the Social Security tax. Since most American workers do not approach the $60,000 level, that means that they no longer have a "Social Security vacation" period late in the year, when Social Security taxes are no longer taken from their paychecks.

In 1994, workers making $60,600 paid about $4,635 in Social Security and Medicare taxes. But their benefits also had been greatly increased over the years. In 1994, the maximum benefit paid a retired worker was $1147 per month. An additional 50% was paid if the worker had a spouse. For some retirees, Social Security is their primary source of income. Were it not for the program, they would be living in poverty.

Conservative writer George Will pointed out in his syndicated column of January 12, 1990, that a rather dramatic change had occurred in the U.S. By the beginning of the 1990s, nearly three fourths (74%) of all U.S. taxpayers were paying more in Social Security taxes than in Federal income taxes.[73]

Some workers have a tendency to think only of the taxes they pay. They forget that sharp increases in the Social Security payroll taxes also dramatically increase employers' labor costs. This amounts, some congressmen say, to an "Anti-Employment Tax" encouraging business to substitute machinery for people whenever possible. It also discourages them from hiring additional workers.

Some economists predicted that large corporations would try to offset increased costs by hiking prices. They also have predicted that some small businesses may be forced to close.

About 40 million Americans get Social Security checks of one sort or another each month. Many of them believe that the system is an "insurance program" in which they pay in funds during their working years and receive benefit checks during retirement as a matter of right.

Such is not the case. Economists call Social Security an *"intergenerational transfer program."* This complex term means simply that today's working people pay taxes to support those now retired or disabled. Today's wage earners ultimately must depend on the willingness of their children and grandchildren to continue these transfer payments.

In 1965, Congress enacted, at President Johnson's request, a *Medicare* program which pays most of the hospital costs of citizens 65 years old and above. Under another (Part B) voluntary plan, it also paid most of their medical expenses.

In 1972, Congress *"indexed"* Social Security benefits, placing them on a par with the Consumer Price Index (CPI). The idea behind this was to prevent benefits from being eroded by inflation. Whenever the CPI rises 3% or more in a given year, benefits are adjusted.

In July, 1978, for example, those receiving benefits got an automatic 6.5% hike. The maximum monthly benefit for a worker retiring in 1977 at age 65 rose from about $460 to $490. By 1994, it was nearly $1150.

The average monthly Social Security benefits for a retired couple rose from $407 to $433 in 1977. By 1994, it was about $675. The inflationary pressures which caused adjustments in benefit checks also caused a major erosion of Social Security trust funds.

From 1937-1975, total taxes going into Social Security trust fund totaled $586 billion. Payments during the same period were $542 billion, leaving a $44 billion surplus to be invested in U.S. securities.

In 1975, the system paid out more than it took in. Some $69.2 billion in benefits were paid while only $66.7 billion in taxes and interest on bonds were received. Looking ahead, the fund appeared to be in serious trouble. Some economists forecast emptying of the trust fund by 1982 and a deficit of nearly $15 billion by 1983.

Congress responded in December, 1977, by enacting the largest peacetime tax increase in American history. The Social Security financing bill hit low and moderate income groups particularly hard and produced an enormous political backlash.

***

XXXIV. THE FUTURE OF SOCIAL SECURITY

"I feel I have been personally burglarized by my own country all in the name of social equality."

"I think it stinks. I have no choice in the matter whatsoever. I gotta pay you all my working life - then I gotta take what you say I can have for what few years, if any, that I live after I'm no longer able to work."

"I strongly believe working wives should receive their own pension in addition to that allowed wives who never have worked...I resent the fact that wives who have never contributed are entitled to a pension that may equal mine. I somehow feel that I have been forced to donate my money."

These three letters were written by people concerned about the United States. They were released to the press as an independent panel began to review the Social Security system in 1978.

Congress has provided, by law, that a non-governmental group study the system at least every four years.

Many students are well aware that Social Security has been the target of periodic criticism and demands for change.

The Social Security debate was intensified by the December, 1977, increase in the payroll tax - the largest ever imposed by Congress. Taxes for highest paid workers tripled by 1987.

Some congressmen, faced with election year battles to save their political hides, urged a reduction in the Social Security tax. But major increases in taxes did not occur until January, 1979, two months after many congressmen assumed that they would be comfortably re-elected.

This time, however, voters reacted. Adding the Social Security tax and proposed energy tax increases, they discovered they would not really get a "tax cut" Congress was planning.

Despite the House Democratic caucus resolution of April, 1977, adopted by a vote of 150-57 urging the use of general tax revenues to finance some Social Security benefits, the House Ways & Means Committee voted to block the election year effort to roll back the Social Security tax increase.

Committee Chairman Al Ullman, (D. Ore.), said: "I have decided that I cannot live with a vote to undo the tax increase." Ullman and the committee may have been motivated by the inescapable facts of demographics.

The population of the United States is proportionally older today than when Social Security was established in 1935. Only 6% of the people were 65 or older then. In the middle 1990s nearly 13% are 65 or older. By 2060, according to government estimates, 23% of the U.S. population will be 65 years of age or older.[74]

As a result of the *aging of America*, the *ratio* between workers who contribute to Social Security and pensioners who take money out of the system is worsening.

In 1977, there were about 30 pensioners for every 100 workers. Demographers estimate that by the year 2050 there may be 50 pensioners for every 100 workers.

With *life expectancy* rates growing, fewer and fewer workers will have to support more and more pensioners.

Although then unpopular, the new Social Security rates enacted by Congress in 1977 infused new money into Social Security trust funds.

The system, Carter administration spokesmen said, was on an even financial keel until the year 2010.

That projection, however, appeared in the early 1980s to be entirely too optimistic. Social Security actuaries believed that the fiscal integrity of the program was threatened by the exponential growth of the system. Some questioned whether Social Security could remain self-financing. Pay-as-you go had been central to financing the program for nearly 50 years. Mills' 1972 indexing legacy had caused a 20% increase in benefit payments.

Demographers, looking ahead, worry about the number of *"Baby Boomers,"* those children born during and shortly after World War II. When they retire, the system might collapse, some pessimists concluded. Clearly, the *dependency ratio* was changing. The ratio measures the number of workers contributing to the Social Security fund versus the number of retirees claiming benefits. There were more "Boomers" than there had been retirees from the previous generation. Even worse, from a Social Security system perspective, the "Boomers" had small families with comparatively fewer children whose contributions would support their benefits when they retired.

The Greenspan Commission

To meet this crisis, President Reagan established a bipartisan commission under the direction of Alan Greenspan to restore the fiscal health of the system. The Greenspan Commission laid the groundwork for subsequent congressional action. The *Washington Post* described its set of proposals as being "as close to absolute fairness as any Social Security revision can ever be."

It did, however, take Congress two years to agree to modest adjustments to the Social Security program. Sen. Daniel P. Moynihan (D., N.Y.) credits Republican Senate leader Bob Dole and James Baker, then White House chief-of-staff, with persuading President Reagan to press for the reforms.

On April 25, Reagan wrote to Senator Moynihan:[75]

"This law demonstrates for all time our Nation's ironclad commitment to Social Security. It assures the elderly that America will always keep the promises made in troubled times a half of a century ago. It assures those who are still working that they, too, have a pact with the future and that they will receive their fair share of benefits when they retire. This compromise proves that bipartisanship can resolve serious national problems. It is a clear and dramatic demonstration of how effectively our system works when men and women join together for the common good. The Social Security Amendments of 1983 are a monument to the spirit of compassion and commitment that unites us as a people."

The amendments, if modest, did have profound consequences. The age of eligibility for benefits was raised, so that today's collegians must wait until 67, rather than 65, to retire on full benefits. Federal and nonprofit employees were brought into the program, making coverage (and contributions) nearly universal.

The *Greenspan Commission* and the Congress hoped to generate large surpluses in anticipation of the retirements of the Baby Boom generation. They also spelled the end of the pay-as-you-go system, placing Social Security financing on a reserve basis.

This time, American taxpayers were told, Social Security would be solvent well into the 21st Century. Robert Ball, former Social Security commissioner, commented on the 50th anniversary of the program:[76]

"Social Security is built on traditional values and concepts - self help, mutual aid, insurance and incentives to work and save. The founding fathers of Social Security planned well, and we are reaping the benefits of their work 50 years later through a program now soundly financed for both the short and the long run."

Some economists continue to worry a good deal about young people and Social Security. One of them, Professor Michael Wachter of the University of Pennsylvania, says:

"Government ought to be preparing the younger generation to face the fact that the Social Security system won't pay for them. But they can't because then people would not go on paying the higher premiums."

Professor Wachter believes that the Social Security system is doomed by the children of the high birthrate 1950s and early 1960s who will eventually have to be supported by the children of the low birthrate decades that followed.

Wachter's view is decidedly pessimistic. Others also are skeptical, not because of numbers, but rather because they are concerned primarily about the lack of political leadership on the part of future Presidents and members of Congress.

In a study done under the auspices of the American Enterprise Institute, a Conservative think-tank, Carolyn Weaver says that the surplus is huge. As a result of the 1983 amendments, a surplus of $1.6 billion a week was being generated and this was expected to reach $1 TRILLION in 1996. By 2030, Weaver says, the surplus should rise to $12 trillion.

However, Social Security reserves are invested entirely in U.S. government bonds and the growing and enormous Federal debt leads Weaver and others to worry that Congress may be tempted to use the surplus, not for future retirement needs, but to fund programs of current consumption.

The Social Security surplus, she reasons, *exists only on paper*, an IOU from the Treasury Department which withholds tax revenues. In theory the interest generated by the surplus will be used to benefit "Boomer" retirements. Because the U.S. counts Social Security withholding as revenue, it offsets the deficit. In short, the Federal deficit is greater in the amount of the Social Security surplus in any given year. In sum, Weaver concludes, the long term health of the program depends on the overall health of America's political economy and American society.

Social Security is now the largest single program in the Federal budget and reached more than $300 billion in fiscal year 1993. In 1995, it became an independent Federal agency and no longer is part of the Department of Health and Human Services.

As wage earners and employers pay high Social Security taxes each year, the program probably will continue to be one of the nation's most controversial social welfare measures.

By July, 1999, Washington, in the unaccustomed position of dealing with a budget "surplus," was engaged in a political debate about how to spend it. President Clinton proposed to use most of the money to ensure the future solvency of Social Security and Medicare. The system faces demographic problems when his large "Boomer" generation inevitably retires, grows old and encounters higher medical bills.

Congress was split along ideological lines. Most Democrats and some Liberal Republicans wanted to develop a bipartisan measure (i.e., "cut a deal") which could be signed by the President.

Some of the more Conservative Republicans in Congress insisted that the "surplus" was the result of the American people being "overtaxed," and that they deserved a tax cut in the range of up to $1 trillion over a decade.

It remained to be seen whether a compromise could be reached or if the parties would harden positions and engage in a bitter partisan struggle during the year 2000 presidential and congressional elections.

***

XXXV. TAX REFORM

What is a good tax?
Some cynics have suggested that it is a contradiction in terms, or something that someone else must pay.

It was more than 200 years ago that Benjamin Franklin said that "in this world nothing is certain but death and taxes." Thanks to the progress of modern medicine, the inevitability of death has been postponed for most Americans until "old age." Most seniors live to see their grandchildren, a blessing not always enjoyed by prior generations. But taxes remain a topic of great prominence, if not an obsession, in debates about America's political economy.

Political and economic analysts long have argued about the tax structure. At the national level, the principle of *"progressive"* taxation has been the foundation of tax policy since adoption of the income tax in 1912.

When Congress considers tax legislation, it is deciding how resources shall be taken from individual and corporate taxpayers to finance various Federal programs. But it also must consider the impact of taxes in redistributing income.

Some economists suggest that people should be taxed according to the benefits they receive from particular public services. This can be simple when motorists are asked to pay tolls or gasoline taxes for improved highways.

But how can one apply the *benefit principle* to public education or national defense?

Should parents who send their children to public schools pay higher taxes to support them than a childless couple? Should single persons and those who send their children to private schools be exempted from public school taxes?

Should the latter get tax credits for educating their children in private schools? These questions are not always easily answered.

Those who believe that education benefits society, rather than the individual alone, argue that all taxpayers have a stake in supporting schools.

They do not stress the *benefit principle* so much as the *sacrifice principle.* Taxpayers, they say, should sacrifice some of their dollars to improve society's educational level.

The principle of *"ability to pay"* has been the basis of U.S. tax laws for many years. Some consider this common sense, a "you can't squeeze blood from a turnip" viewpoint.

Some Liberals, such as John Kenneth Galbraith, argue that because wealthier taxpayers have greater "ability to pay" than middle or lower income groups, their "sacrifice" is less.

Congress and the President, in considering tax laws, must deal with the perplexing question, "How much difference should there be in tax levels?"

Few Liberal congressmen, for example, supported Galbraith's plea in 1974 that they should impose a surtax of 10% on incomes of more than $15,000 and of 20% on salaries of more than $20,000. Single taxpayers with incomes of $32,000 could have paid up to 70% of their money in tax bills, a situation Congress could not seriously consider.

Some ardent "redistributionists" believe that the "rich" should pay higher taxes because they can cope with inflation.

But many families in the $20,000 income range were making mortgage payments, educating their children and trying to pay sharply rising food bills. A *"progressive"* U.S. income tax means that wage earners pay not only more absolute dollars as they climb the economic ladder. They also pay a higher percentage of that income in taxes.

Progressive income taxes also are used by a number of states, such as Ohio and Wisconsin. Other states, such as Michigan, use a flat rate.

"How progressive?" Or "what is the flat rate?" are critical questions.

Income taxes are *"direct,"* levied on people, and politically sensitive. Other types of taxes are *"indirect,"* levied on goods and services. Examples include sales taxes or stamp taxes on cigarettes or liquor.

Property taxes, the traditional source of revenue for local schools, can also be *regressive.* They usually are based on the value of the home, rather than the current economic condition of the homeowners.

Some states, however, do give tax breaks to retirees and others on pensions. Nonetheless, regressive taxes can hit those least able to pay them very hard.

Prior to adoption of property tax reform in Michigan, it was common for retired, elderly couples to be forced out of the homes they had lived in most of their lives simply because they could no longer afford to pay property taxes on them.

Large cities commonly have municipal income taxes.

When these are added to all other taxes, the net result may produce a mood of taxpayer revolt. The Gallup Poll has found that voters are in a Conservative mood on taxes. They favor, it found, a Federal constitutional amendment, requiring government to balance the budget.

Taxes provide revenue for essential government operations. But they also are important, particularly at the Federal level, to promote a stable economy.

Fiscal policy is used to "slow down" an overheated economy when inflation imperils the value of the dollar. The Vietnam era tax surcharge was intended to keep the lid on prices.

Tax cuts, on the other hand, pump money into the consumer's pocket. This stimulates the economy during recession.

During his campaign for the presidency in 1976, candidate Jimmy Carter often spoke of the need for tax reform. But he rarely spoke in specifics, preferring to make a generalized attack on the existing tax system.

A President, however, must be specific. In his tax message to Congress in January, 1977, Carter outlined a series of proposals which sparked debate in Congress, the press, corporate boardrooms, university classrooms and elsewhere.

He proposed:[77]

- A $23.5 billion reduction in personal income taxes, offset some $8.4 billion by tax reforms.
- An $8.4 billion reduction in business taxes, offset $1.1 billion by tax reforms.
- Elimination of excise taxes on telephone and telegraph service and reduction of the Federal unemployment compensation payroll tax.

Carter claimed that the $24.5 billion tax reduction would stimulate the economy, add a million workers to the nation's payrolls and reduce unemployment to 5.5% by the end of 1979. He also said that his tax bill would make the system more "progressive."

A "typical" family of four, earning $15,000 a year, would receive a 19% tax cut, the President said. Families with incomes of less than $15,000 would benefit even more, while those with incomes of more than $15,000 would benefit less.

Treasury Secretary W. Michael Blumenthal estimated that the President's plan would eliminate about six million taxpayers with low incomes from the tax rolls.

A family of four with income of $10,000 would have its tax bill reduced from $446 to $134, according to Blumenthal. A four-person family earning $40,000 would have its tax bill cut from $6848 to $6630, the secretary said.

Few congressmen opposed the idea of an election year tax cut. But many doubted seriously that Carter's plan would do much to reduce the tax bite.

They pointed, in particular, to two factors. A large increase in Social Security payroll taxes already has been mandated for future years and steady inflationary pressures have pushed many families into higher tax brackets, although their standards of living have declined.

The latter factor caused some congressmen to propose "indexing" of the tax system. This would make automatic adjustments in taxes to offset the impact of inflation.

Rep. Willis D. Gradison Jr. (R., Ohio), said:

"I believe strongly that the Federal government should not be the beneficiary of tax hikes that Congress does not levy...The indexation of our tax system is an important step towards insulating all taxpayers against the effects of inflation."

The administration objected to this. Charles Schultze, chairman of the President's Council of Economic Advisers, testified against indexing while supporting Carter's program in Congress.

The President's "tax reforms" sparked more intense debate than his proposed reductions. Carter proposed changes in itemized deductions which would, he said, cause six million more Americans to take the standard deduction, rather than itemize. This would gain $7.8 billion for the Treasury and increase the percentage of taxpayers taking the standard deduction from 77% to 84%, Carter said. Specifically, the President asked Congress to eliminate deductions for: general sales tax, taxes on personal property (real estate excepted) and gasoline taxes.

He also urged Congress to crack down on business entertainment "to curtail inappropriate subsidies, special privileges, inequalities and abuses in the tax system." This could save the taxpayers $1.5 billion in 1979, the President said.

"One individual wrote off the cost of business lunches 338 days of the year at an average cost far exceeding $20 for each lunch. But there is no deduction in the tax laws for the factory workers...or the secretary's lunch with fellow workers," the President said.

Treasury Secretary Blumenthal conceded that the effect of this proposal might be to reduce employment in the restaurant business by about 1%. Studies indicate that the typical lunch consists of a sandwich and soup, rather than "three martinis," but his proposal made headlines for months.

The President, in the American political system, proposes. But Congress disposes. Carter did not get many of his proposals translated into law.

The size of the tax cut was less than originally sought. Carter himself scaled down his request to $20 billion, effective in January, 1979, rather than October 1, 1978, as originally proposed.

Congress also considered its own tax proposals, some of which were fiercely opposed by the Carter administration. The President actively opposed tuition tax credits for non-public schools. A veto battle between Carter and Congress occurred on this issue.

Congress seemed attuned, according to *Congressional Quarterly*, to the plight of the middle class, perhaps more so than the President.

Rep. Clarence J. Brown (R., Ohio) said:

"I've been telling my constituents that you're in the middle-class if you're promised a tax cut, but end up paying the same or more because of higher Social Security taxes."

Congress also took up a reduction in capital gains tax rates.

The administration opposed a proposal of Rep. William A. Steiger (R., Wis.) to roll back capital gains taxes to 1969 levels of a maximum 25%. Capital gains were taxed as high as 49%, which some congressmen felt was too steep to encourage investment.

Carter apparently believed that tax reform was a popular political issue. Government rules are complex and sometimes irritating to taxpayers. But the problem of tax reform is complex. Congress must consider the impact of eliminating exemptions on the economy. Tax incentives have a purpose, such as encouraging home ownership or stimulating industrial production.

The Reagan-Bush View

The election of 1980 was, of course, to set the nation on an entirely new course as far as tax policy was concerned. It was to usher in 12 years of the Reagan-Bush philosophy that taxes were too high and must be cut to restore incentive, that the number of brackets had to be reduced, that the system had to be simplified.

Changes in tax laws in 1981 and 1986 simplified the code, reduced the number of brackets and sharply reduced the amount of taxes that wealthier Americans paid. Conservatives hailed the Reagan Administration for its efforts to create fairness out of the messy Federal tax laws.

Despite considerable fanfare about "closing loopholes" for the rich, the wealthiest 1% of Americans enjoyed a larger share of national income than the poorest 40%. The middle class actually experienced a slight increase from 19.6% to 20.3% in its tax rate, suggesting that "reforms" made have been less useful to moderate income families than the Reagan administration had promised.

A Flat Tax?

During his unsuccessful race for the Democratic Party's presidential nomination in 1992, former California Governor Jerry Brown advocated a flat tax. Quite simply, the flat tax takes the same percentage of taxes from all. It is easy to understand. Brown would have pegged the flat tax rate at 13% of what each individual earned.[78]

Wealthy taxpayers, who previously had paid very little because of endless loopholes, would now pay 13%. However, poor people at the very bottom of the socio-economic ladder also would pay 13%. Some of them had paid no taxes at all under existing law.

Brown's supporters called his idea bold, if not visionary. The plan was direct and understandable. It also would be less expensive to most taxpayers by broadening the base to include everyone.

Brown's opponents argued that the wealthy could always find a way to circumvent the tax, leaving the poor and middle class to bear the burden. It would be particularly hard on those who previously had paid no taxes because of very low incomes.

A flat tax violates the *principle of progressive taxation* by which the rich pay proportionately higher taxes than the poor. Ability to pay long has been recognized as the American standard of fair taxation and government uses tax law not only to raise money but also to redistribute wealth and promote social and economic equality.

Those who thought that the flat tax idea was dead were perhaps surprised at its appeal during the early months of 1996. Steve Forbes, a multi-millionaire publisher, dusted off the concept and won considerable support for it from some Republican presidential primary voters and others.

Main features of Forbes proposal:[79]

(1) One tax rate of 17% to replace the current five rates (brackets). He would allow exemptions for singles at $12,800 and for couples at $25,600 plus $5200 for each dependent.

(2) The plan would end deductions for mortgages, interest, state and local taxes and charitable giving.

(3) Interest, dividends and capital gains would be exempted from taxes. He would end the inheritance tax and those on Social Security benefits.

(4) The corporate tax rate would be cut from 35% to 17%, but they no longer would be able to deduct payments for fringe benefits (other than pensions), interest and the business half of Social Security withholding.

Forbes became known as a one issue candidate and was forced to withdraw as Senator Dole left him far behind in the GOP race. But Forbes threw his "hat in the ring" again in 1999, seeking the presidential nomination for 2000.

***

XXXVI. THE ENERGY CRISIS

Most students entering college this fall, if asked to identify the most pressing economic problem facing the U.S., probably would not even think of the energy "crisis."

Yet in his first appearance before Congress as President, Jimmy Carter declared in April, 1977, that the energy problem was "the greatest domestic challenge that our nation will face in our lifetime."

More than a year later, despite two of the most devastating winters in U.S. history, a comprehensive energy program was moving only slowly through Congress.

Many Americans need not be reminded of the impact of the winters of 1977 and 1978 on their lives. Natural gas shortages forced closing of industries and schools and created serious unemployment in several states.

Utility bill increases contributed significantly to inflationary pressures. So did gasoline price increases following the six-month oil embargo that followed the 1973 Arab-Israeli war.

What were the roots of the energy problem? What "solutions" were proposed to deal with it? Can the energy problem be resolved within the context of our traditional democratic values and free market economy?

These were merely a few of the tough questions arising during the Carter Administration. Past practices, however, were at the root of the energy problem.

Historically, Americans have had tremendous appetites for energy. They sometimes have behaved as if energy supplies are inexhaustible. Economic growth and improved standards of living, traditional American values, have resulted in 5% of the world's population using about 33% of the world's energy supply.

Of vital significance is the kind of energy we have used. We have relied on fossil fuels. Oil, a finite resource, has supplied nearly half of our energy supplies. Natural gas, which some believe will last only another 30-40 years, accounts for about a third of U.S. energy needs.

Coal, which may last for several hundred more years, supplies about 20% of our needs.

Hydroelectric power (less than 4% of energy supply) and nuclear power fulfill the rest of our needs.

Few could have foreseen the present dilemmas at the dawn of the 20th Century. Enormous growth in demand followed Thomas A. Edison's development of electricity for the home. Edison supplied only 59 customers with his first power plant.

Mass production of automobiles had enormous impact on oil firms.

The 1930s and 1940s were decades of rapid growth in both supply and demand. Some charge that Americans became energy "gluttons" in the post-Word War II era. Between 1930 and 1960, there was a ninefold increase in demand for electric power.

New technology made life more convenient for the post World War II generation. A television commercial of the time featured the "all electric girl." She glowingly portrayed the advantages of an all-electric kitchen.

Americans enjoyed not only the benefits of radios and television sets, but also power mowers, blenders, skillets, washers and dryers, air conditioners, electric typewriters and even electric toothbrushes. A lifestyle unknown to our grandparents put enormous pressure on energy supplies.

Until 1950, the U.S. was able to supply its own energy needs. But demand was doubling every 12 years. Whereas more than half of U.S. energy came from coal before World War II, postwar changes put heavy demands on other energy sources.

By 1975, coal supplied only 19% of U.S. energy requirements. Crude-oil products and natural gas supplied 75% of energy needs, hydroelectric and nuclear power the rest. The U.S. became more and more dependent on foreign energy sources during the 1960s and 1970s.

For years, government policy made it more attractive for U.S. oil companies to build refineries abroad, rather than enlarge refining capacity at home. Federal Power Commission (FPC) policies, kept prices of natural gas below what the free market would have established. These artificially low prices paid by consumers for natural gas encouraged its use. Less demand was made for heating oil, encouraging oil companies to place what some regard as an "exaggerated importance" on gasoline in their marketing policies.

Ironically, the U.S. became increasingly dependent upon foreign oil precisely when the power of the Organization of Petroleum Exporting Countries (OPEC) grew. In 1960, for example, the United States imported 18% of its oil supplies. By 1972, it imported nearly 30% of its oil.

The period from October, 1973, until March, 1974, was characterized by motorists waiting in long gasoline lines, rapidly rising fuel prices, talk of rationing and frantic government efforts to evolve an energy policy.

In 1975, Congress passed the *Energy Policy and Conservation Act,* authorizing the Federal Energy Administration to order utilities to burn abundant coal rather than expensive oil. The act also created a strategic petroleum reserve as a hedge against future OPEC boycotts. Detroit was ordered to improve energy efficiency of auto engines. The U.S. also encouraged development of nuclear power plants.

The energy crisis produced a recession in the U.S. Hardest-hit were the Northeast and the old industrial centers of the Upper Midwest. These states imported most of their energy. The crisis for "rust belt" cities came in the 1970s and early 1980s. In Michigan a popular bumper sticker of the time - reflecting extreme pessimism about the state's economy - read "Will the Last Person To Leave Michigan Please Turn Out The Lights?" But the move to Texas and other "sunbelt" states was short-lived. In the mid-1990s, Michigan was known as the "comeback state."

Presidents Nixon and Ford set energy "independence" by 1980 as a national goal. The Federal Energy Administration (FEA) was given the job of formulating policies to achieve that goal. But with the end of the oil embargo and disappearance of gasoline lines, people appeared to lose interest in the "crisis." By February, 1978, the U.S. was more dependent on foreign oil than ever before, importing nearly 50% of its supplies.

Some economists question how "independent" the United States really should be. Many perplexing policy questions exist, not the least of which is the cost-benefit calculation.

The United States could become self-sufficient in energy production. But at what cost? National security considerations are of vital importance. But are the American people ready to pay the price for "independence?"

Carter's description in 1977 of the energy problem as the "moral equivalent of war" was not shared by his fellow citizens, according to public opinion researcher George Gallup. The pollster found that 40% of the American people are unaware that the United States must import oil to meet its energy needs.

By the end of the Carter Administration, the "moral equivalent of war" was sounding to many Americans like MEOW, its acronym.

Two days after giving his energy message to the public, President Carter submitted an energy plan to the 95th Congress. He proposed:[80]

(1) An increase in the Federal gasoline tax of five cents per gallon in each year for a period of 10 years, beginning in January, 1977. Eventually domestic oil prices would rise to levels of expensive imported oil. Funds would be rebated to the public through a complex formula.

(2) A tax on "gas-guzzlers." Cars burning more fuel than permitted by U.S. standards would be taxed. The worst (or least fuel efficient) vehicles would draw taxes of $2488 by 1985.

(3) Utility companies would be ordered to offer home insulation financed by loans to be repaid through monthly utility bills.

(4) He recommended tax credits for homeowners (up to a maximum of $410) for insulation. He also proposed tax credits for industrial conversion to coal.

(5) U.S. crude oil would be taxed at the well-head to raise domestic oil prices by 1980 to world levels.

(6) Federal control over natural gas sales would be extended to intrastate gas for new gas production only. Coal use would be encouraged and required in new plants. Tax credits would offset conversion costs.

(7) Licensing of conventional nuclear reactors would be speeded up and safety standards strengthened.

The President, who had taught Sunday School in the Baptist Church at Plains, Ga., also used the White House to preach a bit. Teddy Roosevelt once had called the presidency a "bully pulpit."

Americans, Carter said in one of his homilies, should lower their home thermostats to 65 degrees. They should take only essential auto trips, use car pools and drive no faster than 55 miles per hour to save gasoline. Congress eventually was to mandate lower speed limits as a part of the fuel conservation movement.

Reaction to the President's complex plan was mixed, both on Capitol Hill and among the many affected interests.

Sen. Robert C. Byrd (D., W.Va.), Senate majority leader, said: "The President cannot expect every jot and tittle to be exactly as he proposed it."

Pledging his support to the President, House Speaker Thomas P. (Tip) O'Neill (D., Mass.) said: "This is his first major fight. This is a battle. He's going to have to give us some help along the line."

Thomas Murphy, chairman of General Motors Corp., called the plan to impose a tax on "gas-guzzlers: rash, ill-conceived, ill-prepared...the most simplistic, irresponsible proposal ever made."

James R. Schlesinger, Carter's chief energy adviser, countered, "I guess what is good for General Motors is still not necessarily good for the United States."

Roy Chapin, chairman of American Motors, a firm which emphasized small car production, called Carter's proposal "a sensible approach."

The President's proposals were intended to stress conservation by making consumers pay more for energy. He was using a fundamental principle of economics: *Prices are a rationing device.*

Congress moved with great deliberation on this complex issue. It did establish a *Department of Energy (DOE)*, headed by Schlesinger, to co-ordinate work previously done by a large number of Federal agencies, including the Federal Power Commission, Federal Energy Administration and Energy Research and Development Administration.

In his January 20, 1978, Economic Message to Congress, Carter again urged enactment of a national energy policy to reduce consumption and encourage production. In March, 1978, Senator Russell Long (D., La.) said that the "centerpiece" of Carter's energy tax proposal, the crude oil equalization tax, was dead.

Earlier, congressional committees had killed other aspects of the President's plan, including his proposal to rebate increased Federal gasoline taxes to the public.

Congress was slow to respond to proposals for increased taxes in the midst of a "taxpayer rebellion."

After complaining about congressional rebuffs to his energy plan, Carter began to express some willingness to compromise.

In May, 1978, House-Senate conferees approved a compromise plan for natural gas pricing, breaking the energy logjam. They agreed to lift U.S. price controls on domestically produced, newly-discovered gas on January 1, 1985, and to provide for a 10% annual increase in gas prices until then.

Given the frigid winters of 1976-77 and 1977-78, Congress was jolted by energy politics. The 95th Congress produced a weak *National Energy Act*. The measure, very different from what Carter recommended, provided tax credits for installation of energy savings equipment, encouraged utilities to convert to coal, and permitted price increases for newly discovered gas. The law left the U.S. ill-prepared to deal with the dislocation of international oil markets after the Iranian Revolution of 1978. Congress preferred halfhearted measures that left America's future in the hands of the OPEC cartel.

Many Conservatives argue that an energy crisis is the result of not following a "full cost" pricing policy. Given a choice between economic growth and environmentalist demands, they tend to prefer growth. Historically, they argued, the American Way is to produce more, not live with less. Liberals have urged more government regulation. Some of them question whether the free market system can solve the problem.

Some Americans thought that the answer to the "Energy Crisis" was nuclear power. But a heated national debate followed an incident at Three Mile Island nuclear power plant near Harrisburg, Pa.

Environmentalists and others argued that nuclear plants were constructed without adequate care being given to safeguards. Another troubling question was: "Where are we going to dispose of nuclear radioactive wastes produced by these plants?"

Economist Robert B. Carson says in *Economic Issues Today*, "The market system did not invent pollution and interference with the ecological balance...A visitor to many Russian industrial centers will probably note that the air quality of Donora, Pa., is better."

***

XXXVII. CAPITALISM AND FREEDOM

As Americans debate the wisdom of particular economic policies in the latter part of the 20th Century, some consider the large question: "Is capitalism itself the root problem of the American system?"

Can problems of the modern economy - energy, inflation, monetary and fiscal policy, unemployment, and economic growth and stability - be resolved within the context of the existing economic system?

Will values such as individual pursuit of self-interest and profit, consumer choice and efficiency endure?

Socialist critics traditionally maintained that capitalism is doomed. The late Soviet premier and Communist Party leader Nikita S. Khrushchev boasted in 1959 when touring the U.S., "Your grandchildren will live under socialism."

Most Americans considered such a boast a typical example of Khrushchev's tendency to exaggerate what he and his Bolshevik brothers could achieve.

Defenders of the American system quickly pointed out that some problems may appear inherent in capitalism but that others, equally or perhaps more serious, were inherent in socialism. In particular, these problems grew out of centralized, bureaucratic planning and unwieldy state economic organization.

Indeed, socialist theoretician Oskar Lang wrote some 40 years ago that "the real danger of socialism is that of the bureaucratization of economic life."

One of the most articulate defenders of American capitalism is Professor Milton Friedman, former University of Chicago economist and noted conservative writer. Friedman, like many other economists, is greatly concerned about human freedom, as well as efficiency of production. In his landmark work, *Capitalism and Freedom,* he wrote:[81]

"The kind of economic organization that provides economic freedom directly, namely competitive capitalism, also promotes political freedom because it separates economic power from political power and in this way enables the one to offset the other."

Friedman views the growth of big government as the greatest threat to economic progress. His view of the limited role of government is clear:[82]

"A government which maintained law and order, defined property rights, served as a means whereby we could modify property rights and other rules of the economic game, adjudicated disputes about the interpretation of the rules, enforced contracts, promoted competition, provided a monetary framework, engaged in activities to counter technical monopolies and to overcome neighborhood effects widely regarded as sufficiently important to justify government intervention, and which supplemented private charity and private family in protecting the irresponsible...would clearly have important functions to perform. We take freedom of the individual...as the ultimate goal in judging social arguments...In a society freedom has nothing to say about what an individual does with his freedom. It is not an all-embracing ethic. Indeed, the major aim of this is to leave the ethical problem for the individual to wrestle with."

Some critics of late 20th Century capitalism argue that free enterprise philosophy is an anachronism. It may have made sense, they argue, in the era of small firms and free competition. But today large corporations have an enormous social impact. They must either assume social responsibilities and be sensitive to public attitudes or lose their essential autonomy, they argue.

But Friedman and others disagree. What is the public interest? Is it simply what the President says? Is the public interest necessarily superior to private interests or desires. Friedman says:[83]

"Few trends could so thoroughly undermine the very foundation of our free society as the acceptance by corporate officials of a social responsibility other than to make as much money for their stockholders as possible...Can self-selected private individuals decide what the social interest is? ...If businessmen are civil servants rather than the employees of their stockholders, then in a democracy, they will sooner or later be chosen by...election...long before this occurs, their decision-making power will have been taken away from them."

Liberal economists of the "left" disagree. Heilbroner, for example, sees a basic crisis in capitalism. "Profits in corporate irresponsibility," he says, "are business atrocities like My Lai."

Indeed Heilbroner and Galbraith, among others, seriously question whether capitalism - as it presently exists - can survive in the long run.

***

XXXVIII. RONALD REAGAN AND THE SUPPLY-SIDE "REVOLUTION"

When Ronald Reagan ran for President in 1980, he stressed many traditional Republican themes. He attacked Democrats for producing unbalanced budgets and, from a Conservative perspective, criticized their lack of "fiscal responsibility." He spoke of the large Federal debt, then about $910 billion. But perhaps most of all, he spoke of the tax system. Cuts were urgently needed, he said, to restore incentive and productivity in America.

Reagan had majored in economics while a student at Eureka College in Illinois. He fancied himself something of an expert in the discipline, although what was taught was *pre-Keynesian*.

In his book, *America in Search of Itself,* Theodore White writes:[84]

"What a man majors in at college makes him, in his own mind, forever after an authority on that subject...Economics was taught simply: supply and demand, the relation of free enterprise to government, the economic cycle as defined before the New Deal undertook to flatten the curves."

Few adults who watched the last of the Carter-Reagan presidential debates in 1980 will forget Reagan's closing remarks:[85]

"Are you better off than you were four years ago? Is it easier for you to go and buy things in the stores than it was four years ago?...Is America as respected throughout the world as it was?...I would like to lead that crusade...to take government off the backs of the great people of this country and turn you loose again to do those things that I know you can do so well, because you did them and made this country great."

In February, 1981, Reagan, sent to Congress his Program for Economic Recovery. It outlined the main elements of his policies, quickly dubbed "Reaganomics" by the media. What was "Reaganomics?"

It was, in part, a program based on the view that government programs, particularly Federal government programs, were likely to be wasteful and inefficient. Consequently, the new Chief Executive wanted to shrink the rate of growth in Federal spending. This, however, was easier said that done since about 70% of the Federal budget is *"uncontrollable."*

The late Pulitzer Prize winning journalist-historian Theodore White described the problem in *America in Search of Itself* as follows:[86]

"The reality is that no one, absolutely no one, can control the budget of the United States - not the President - not the OMB (Office of Management and Budget) - not Congress. The moral mandate of the sixties and seventies locks expenditures into an ever-rising, irreversible escalation...Ever-rising interest on the national debt must be paid...This is the reality."

Most of the U.S. budget consists of things like Social Security, Medicare, various other "entitlement" programs, most of which were enacted during Lyndon Johnson's "Great Society," and interest payments on the national debt.

Since President Reagan was dedicated to the proposition that the U.S. must never again suffer the kind of humiliation it did during the Iranian hostage crisis, curtailing military spending was out of the question.

The Reagan years, in fact, were characterized by a massive U.S. arms buildup. The President felt that there had been a unilateral arms race in which the Soviet Union had moved ahead of the U.S. He was in no hurry to negotiate with the Soviets until the U.S. had improved its national security position.

Therefore, all budget cuts would have to come out of domestic civilian programs. Unfortunately for budget cutting enthusiasts, this constituted only about one-eighth of the total budget.

Economic interest groups and their congressional supporters also were dedicated to protecting their positions. Politically, big labor, big business and big agriculture were too important to be "maltreated."

The President also advocated curtailed spending on "welfare." But what is welfare? Americans, according to public opinion polls, are not very sympathetic to the idea of "welfare," but semantics in survey research is everything. What words are used and how questions are asked is of vital importance. When pollsters ask questions about "relief," public opinion shifts in a more compassionate direction.

Soon the press was using the term *"safety net"* to refer to the "deserving poor." The American people, surveys have shown, believe that unfortunate people who are victims of circumstances beyond their control must be helped to escape from their misery.

Most of the American media do not cover the intricacies of economic policy. But talk of the "Reagan Revolution" and the new team in Washington led to a proliferation of stories about *"supply-side"* economics.

Supply-side economics was a reaction to the problems of the 1970s and what many saw as the excessive influence of Keynesian fiscal policy ideas. When Reagan succeeded Jimmy Carter as President in January, 1981, the U.S. was suffering from stagflation - the worst of both economic worlds.

Simultaneously, the U.S. was experiencing high unemployment (10.5%) and high inflation (13.9%). Keynesian theorists long had taught that rising inflation triggers reduced unemployment and visa versa. When, however, both conditions existed at the same time, many politicians, as well as economists, lost their appetites for Keynesianism - which failed to anticipate, much less cope with stagflation.

Supply-siders advocated a move away from Keynesian theory and a return to the ideas of *laissez-faire* and Adam Smith's "invisible hand." Political Conservatives were quick to adopt this approach. For them, the theory offered a hope of simultaneously reducing unemployment and inflation while also decreasing government spending and producing a balanced budget. Of paramount importance, all of this could be done, they believed, with less - not more - government involvement in the economy, a long-cherished conservative economic principle.

The key to supply-side economics, as its name suggests, is its focus on supply in the economy, rather than on demand, as in Keynesian economics. Keynes had suggested that, in order to combat inflation, governments should try to cool off the economy by reducing demand for goods and services. Supply-side theorists suggested that inflation could be lowered by increasing supply.

Supply-siders emphasized the importance of the private business sector. Giving incentives to business to invest and to expand production, thereby to stimulate supply is, according to the theory, the key to reducing inflation and stimulating the economy.

Democratic Party Liberals were unimpressed. Supply-side economics, they said, was nothing more than a "wolf in sheep's clothing," a warmed-over version of "trickle down economics," popularized during the Coolidge-Hoover administrations.

Taxes and Incentives (or Disincentives)

Reagan believed, in light of his own personal experience as an actor, that both individual and corporate income taxes were too high. They created a problem of disincentive in the economy. He often related stories of his days in Hollywood. He refused to make more than two films a year because most of the additional income that he could have earned would have been taken by taxes. Productivity, he reasoned, is reduced by high tax rates.

Reagan was convinced that *marginal rates* of taxation must be reduced. The marginal rate is that at which additional income is taxed. Before the Reagan Administration, the marginal rates of the Federal personal income tax ranged from 14% to 70%. Supply-side economists argued that these high marginal rates, particularly those in the 50%-70% range, reduced productivity.

If, they reasoned, a person must pay 70% of his income to the government in the form of taxes, he or she will prefer leisure time over additional work. If one knows that his or her "risk capital," if successfully invested, will yield a 70% tax return to Uncle Sam, where is the incentive to invest?

Supply-siders also argued that high marginal tax rates encouraged individuals to seek out *"tax shelters,"* which are unproductive investments favored by tax laws that reduce personal income taxes. A large "underground economy" flourishes when marginal rates are high and people are encouraged to trade goods and services rather than conduct transactions in the open where they will be subject to taxation.

Arthur Laffer, a young economist at the University of Southern California, argued that higher tax rates do not produce the higher revenues they are expected to provide because of these disincentives. The *Laffer Curve* depicts a relationship between tax rates and tax revenues. As rates rise from zero to 100%, revenues increase from zero to some maximum level and then decline to zero. Tax revenues decline beyond a certain point, Laffer said, because higher rates discourage economic activity and diminish the tax base. Laffer and other supply-siders believed that if America wanted to be made more productive, people would need incentives to work. Therefore, tax cuts were a matter of top priority. Corporate tax cuts, coupled with decreased government regulation of business, would bring greater production and large-scale expansion.

Supply-siders assumed that the beneficiaries of these new economic policies actually would turn a significant percentage of their after-tax profits into new factories, new research and development and new jobs. All of this would occur and those already gainfully employed would enjoy higher incomes.

It also was argued that there would be less tax evasion and tax avoidance because lower tax rates would reduce the motivation to engage in such things.

This new economic approach was an interesting and fundamentally different approach to U.S. economic policy-making, compared to what had gone before for nearly half a century.

Economic Recovery Act of 1981

Reagan tried to combine this *laissez-faire* approach with a substantial tax cut to stimulate investment. The first of the Reagan tax cuts came in 1981. The *Economic Recovery Tax Cut Act of 1981* reduced marginal rates from a range of 14% to 70% to a range of 11% to 50%. During his second term, Mr. Reagan persuaded Congress to enact the *Tax Simplification Act of 1986,* which further reduced rates to only two brackets, 15% and 28%. Thus, top marginal rates came down from a high of 70% to a high of 28% during the Reagan presidency.

Proponents of Reaganomics argued that government revenues actually would be boosted. A prosperous and productive economy would mean an increase in the amount of taxable income. In theory, this would more than offset the tax cuts

The experience of the Reagan years, however, shows that the administration overestimated the stimulative effect the tax cuts would have on Federal revenues. The deficit rose beyond anything imagined. Some of his critics argued that Reagan's legacy may well be the record national debt increase that more than tripled during the 1980s and the soaring budget deficit that continue to this day to plague the U.S. Treasury and tie the hands of the President and Congress as they attempt to manage the economy.

In 1980, the Federal debt stood at $908 billion. It soared to more than $2.87 trillion in 1989. In mid-1996 it stood at about $5 trillion.

Gramm-Rudman

In 1985, three senators, Phil Gramm (R., Tex.), Warren Rudman (R., New Hampshire) and Ernest Hollings (D., S.C.), took the lead as Congress attempted to get more "discipline" into its budget-making process.

A considerable number of voters appeared alarmed by the apparently unending series of unbalanced budgets. The U.S. government failed to raise sufficient revenues or cut enough spending to put its fiscal house in order.

The *Balanced Budget and Deficit Reduction Act of 1985* (popularly known as the Gramm-Rudman-Hollings Act) proposed to shrink the deficit in each of the next five years. Ultimately the budget would be balanced in 1991. Gramm-Rudman also provided a mandatory deficit cutting procedure that came into effect automatically if Congress failed to meet its deficit reduction goals.

Failure to meet prescribed budget targets would be "punished" by a *sequester* of appropriated funds. In effect, this meant that an across-the-board spending cut would be required. But some 48% of the budget items (entitlements plus interest on the national debt) were exempted from the sequester rule.

Another 24% would see only limited application and only 28% of the budget was likely to bear the full brunt of sequesters under Gramm-Rudman.

In 1986, the U.S. Supreme Court declared some provisions of the Gramm-Rudman Act unconstitutional because of technical concerns. Congress then passed virtually the same concept into law a year later, resetting the clock for another five year period. Congress used what reporters called "artful dodging" in constructing the Federal budgets of 1988 and 1989. When Gramm-Rudman was enacted in 1985, its sponsors declared that it was "a bad idea whose time has come." Something, they reasoned, simply had to be done to get the deficit under control. Many citizens had equated the failure of Congress to balance the budget with "financial irresponsibility."

But the "medicine" prescribed by Congress proved to be too strong. It produced an "allergic reaction" among those interest groups whose benefits were threatened. To some members of Congress, the political costs were too high. When Congress failed to discipline itself, a budget crisis occurred in 1990, one which President Bush had to face. For him, the Gramm-Rudman medicine was politically fatal in 1992.

Not once during the five year phased reduction period did the scheduled decline ever come close to its goal. By fiscal year 1992, when the deficit was supposed to be "only" $28 billion, budget analysts projected a shortfall of $362 billion.

Senator Hollings was so angered by congressional failure to balance the budget and reduce the deficit that he announced that he wanted "a divorce from Gramm-Rudman-Hollings." Senator Rudman announced in 1992 that he would not seek another term because of his frustration with deficit reduction gridlock.

Deregulation

Another key element in Reaganomics was deregulation, a move already begun on a small scale by the Ford and Carter administrations. The Civil Aeronautics Board (CAB) was dissolved on January 1, 1985, although such a move had been suggested during the Carter years by Alfred Kahn.[87]

By the end of the 1980s, some economists and others began to argue the case for re-regulation in airline and trucking industries, and particularly in the savings and loan industry after the disasters of the late 1980s.

More than one-third of the 3000 S&Ls existing in 1987 no longer are in business. Without timely action by President Bush and the Democratic-controlled Congress, collapse of the "thrifts" may have threatened the nation's economy. But the Federal government had pledged its "full faith and credit" to back banks and thrift institutions. The final bill to the taxpayers for bailing out the S&L industry is estimated by some at about $250 billion.[88]

Monetarism

The monetarist ideas of economist Milton Friedman got a friendly reception in the Reagan White House. Monetarists believed that careful growth in the money supply would smooth out the undesirable fluctuations of the business cycle more effectively than the Keynesian prescription of more government intervention in the economy. Monetarist policies were embraced by Reagan and Paul Volcker, chairman of the Federal Reserve Board.

Reagan's "Supply-Side Revolution" had counted heavily on dampening the fires of inflation by continuing a policy of very tight money. This appeared to be somewhat at odds with another tenet of supply-siders: spurring the economy on. Tight money policies usually have precisely the opposite effect.

But the war on inflation, clearly one of the main causes of Carter's 1980 defeat, was of primary importance. The rate of inflation fell from 12.4% in 1980 to less than 7% in 1982. Interest rates fell from a record 21.5% early in 1981 - the record high in U.S. history - to 10.5% in 1982.

Perhaps the greatest positive legacy that the Reagan administration left was relatively low rates of inflation and relatively low interest rates.

The New Federalism

Reagan, like many other Conservatives, believed in decentralized government. He was convinced that Washington politicians had usurped many of the traditional functions of state and local government. Like Thomas Jefferson, he believed that the closer the government was to the people, the better. It was reasonable to assume, he often argued, that local people know more about local problems than Washington "bureaucrats."

One way of reducing the Federal budget was to turn over responsibility for some programs to the states. There was a great deal of discussion in the Reagan Administration of the "New Federalism." And, for the first time in 30 years, Federal aid to state and local governments actually declined.

Reagan persuaded Congress to consolidate many grants for specific programs that often require matching funds (*categorical grants*) into far fewer, less restrictive *block grants*. Block grants are given to states for specific purposes, such as health or education, with very few strings attached. In 1981, Congress consolidated some 57 categorical grants into seven block grants. Although happy at first with their new freedom from Federal restrictions, governors and mayors changed their attitudes when they learned that the Reagan administration planned to cut financing of these programs by 25%.

In 1980, the last year of the Carter Administration, Federal funds had made up 26% of state spending and nearly 18% of city and county budgets, respectively.

By 1995, most block grants fell into only four categories: health, income security, education or transportation. Nonetheless, some governors continued to urge consolidation of more programs into block grants. Welfare reform, for example, although the topic of much rhetoric and little action by President Clinton and the Republican-controlled Congress (as of mid-1996) urged more latitude to the states in an effort to get back to the old concept of states as laboratories or centers of experimentation.

Results of supply-side economics were mixed. At first, tax cuts did stimulate productivity and inflation dropped dramatically, from nearly 14% in 1980 to about 4% in 1988.

But although inflationary fires were dampened, a "tradeoff" took place in the form of higher levels of unemployment. In fact, in 1982 (9.7%) and 1983 (9.6%), the rate of joblessness rose to its highest level since the end of World War II. The figures did come down somewhat during Reagan's second term. He left office in 1988 with an unemployment rate that was only 5.5%.

Some Liberal economists also argued that tax cuts on corporations and the wealthy, combined with the largest peacetime military buildup in U.S. history, contributed to the recession after Reagan left office.

XXXIX. GEORGE BUSH, THE ECONOMY AND THE "VISION THING" PROBLEM

As a student at Yale, George Bush had majored in economics and was a Phi Beta Kappa scholar. He knew that there was no such thing as a free lunch. When running against Ronald Reagan in the Republican presidential primaries of 1980, he hit hard at the former California governor's ideas. Theodore White described it as follows: [89]

> "He was convinced that inflation was too complicated to lend itself to one-line solutions his rival Reagan proposed...He spoke well and eloquently about national issues; he drove hard on Reagan's economics. The thought that one could cut taxes and thus increase revenues, all the while jumping the defense budgets to new heights was, he said, 'voodoo' economics."

But after his nomination as Reagan's vice presidential running mate, Bush experienced what some reporters described as a conversion of faith comparable to that of St. Paul on the road to Damascus.

Vice Presidents, of course, simply do not criticize Presidents. Above all else, they must be loyal. During the eight years of the Reagan presidency, Bush accepted Conservative economic and social principles, although how comfortable he was with them is a matter of speculation. Some read into his reference of a need for a "kinder, gentler America" a gratuitous slap at Reaganomics. That, however, is a matter of one's perception.

The election of 1988 is generally considered an endorsement by the American electorate of the *status quo*. The country was enjoying peace and prosperity. So Bush made relatively few changes in the overall direction of economic policy. He asked Congress to cut taxes on capital gains in order to stimulate investment and production and to create jobs. He was opposed to increasing taxes. His New Orleans speech accepting the Republican Party's 1988 presidential nomination, is famous - or infamous, depending on one's point of view. "Read my lips, no new taxes."

Generally, President Bush enjoyed good relations with the Democratic-controlled Congress until May, 1990. He then got into a major battle with leaders of both parties over the projected deficit in the Fiscal Year (FY) 1991 budget. In October, a coalition of House Democrats and Republicans defeated a deficit reduction package that the administration had negotiated with a bipartisan group of congressional leaders at a "budget summit."

Democratic Liberals objected to program reductions. Republican Conservatives complained that the President's "no new taxes" pledge had been broken. Both Liberals and Conservatives were angry that they had been excluded from the group of "insiders" who had overridden the normal budgetary process. The vote in the House was 254-179 against the President. It was a major defeat.

After several weeks of bitter wrangling which included charges on both sides of bad faith on the part of the other, the Bush administration and the Democratic Party-controlled Congress "cut a deal" intended to balance the budget by 1996. The agreement stipulated that for the next three years any reductions in defense or discretionary domestic spending be devoted to reducing the deficit. But unanticipated spending for the Persian Gulf War in 1991, bailing out failed savings and loan institutions insured by the Federal government and shortfalls in projected revenues, combined to produce enormous deficits.

The 1990 budget crisis led to further tinkering with the Gramm-Rudman process. Both Democrats and Republicans, fearing the political fallout from a sequester on the eve of mid-term congressional elections, ultimately made a deal with the President. The agreement meant spending cuts, even in some entitlement programs, and a tax increase.

To prevent another future budget crisis, Congress placed caps on *discretionary spending* - that not required by existing laws - for the years 1991-1995. It also adopted a pay-as-you-go principle requiring that bills providing increases in discretionary spending or decreases in revenues be accompanied by offsetting adjustments elsewhere in the same program category or they could be ruled out of order.

Some Conservatives, including President Bush, argued that Congress simply lacked the will to control spending. The best budget "medicine" would be a *line item veto*. This type of veto, possessed by nearly all state governors, would enable the Chief Executive to reject a particular item in a bill, rather than being forced to accept the entire package. Bills sometimes are cluttered with *"riders,"* irrelevant provisions funding projects in states or districts of members of Congress.

In 1996, Congress gave the President the line-item veto, although it was not scheduled to go into effect until the following year. The Supreme Court declared it unconstitutional in the case of *Clinton v. City of New York*, No. 99-1374 (1998).

Another favorite Conservative idea was to adopt a constitutional amendment requiring Congress to balance the budget each year, except in case of war or national emergency. Although supported by both Presidents Reagan and Bush, this was an idea whose time had not yet come.

It is always politically easier for members of Congress to vote for spending increases rather than program cuts. It is unlikely that an incumbent member of Congress ever was defeated for voting to cut his or her constituents' taxes. The same cannot be said of those who supported tax increases. Some cynics point to members of Congress who favor appropriations bills but oppose taxes to pay for the programs involved. This approach is not only economic insanity. It is the height of political irresponsibility.

But ours is a political economy. What is good economics may be bad politics. What is good politics may be very bad economics.

In 1990, President Bush pursued a course of action called "courageous and responsible" by his supporters. But some Conservative voters were outraged. They charged that Bush was a modern day Judas who had "betrayed" them by violating his 1988 pledge not to raise their taxes.

Time magazine White House correspondents Michael Duffy and Dan Goodgame trace the long friendship of Bush and Dan Rostenkowski, then powerful chairman of the House Ways & Means Committee. "Rosty" urged the President to:[90]

"Tell the people...that if we don't balance our budget, we're going to be number two, and they'll say, 'The hell we will!' If you challenge them, Mr. President, they will accept whatever sacrifice you say is necessary. If you lead, they'll follow."

Bush's reply was half in jest, but it was a harbinger of things to come: "It's easy for you to say."

The President, with great reluctance, agreed to tax increases of $146 billion as a part of a five-year $496 billion deficit reduction package. Top tax rates were increased from 28% to 31% for the wealthiest taxpayers. A number of itemized deductions ("loopholes") also were phased out.

Excise taxes were imposed on gasoline. There were new taxes on beer, wine and liquor and a "luxury tax" on expensive automobiles and yachts. More thought was given to the "high lifestyle" of those who drove the cars or sailed the yachts, than the impact that new taxes and higher prices would have on the jobs of those workers who built these items.

The wealthy bought used yachts and the people who produced expensive watercraft lost their jobs.

Later, during the presidential campaign of 1992, Bush admitted that he had made a "mistake" in going along with the tax hike. The President had tried to explain that 70% of the deficit reduction package came from spending cuts and only 28% from "tax changes," but the memory of the tax hikes alienated economic Conservatives throughout the campaign of 1992. Rather than balancing the budget, as he had pledged to do, Bush saw a fiscal 1991 deficit of about $269 billion and an even higher $290 billion shortfall, the highest in American history in 1992.

In that year, interest on the national debt amounted to more than $292 billion. This was hardly a plus in his campaign for re-election against Arkansas Gov. Bill Clinton and Ross Perot. Both Clinton and Perot hammered away at the President. He was vulnerable on the recession, the rising national debt and other economic issues.

Perot, an independent billionaire financing his own campaign, apparently had a strong dislike for Bush. He went so far, perhaps at the cost of destroying his own credibility, as to charge that the Republicans planned to disrupt his daughter's wedding. Many students of political behavior have concluded that Bush suffered more than Clinton did from Perot's third party candidacy.

Although some evidence supports the view that the U.S. economy had in fact rebounded before the November, 1992, election, voters had a different perception. Some economists say that the recession ended in March, 1991. But the recovery was weak, the pocketbook issue was strong, and the voters concluded that they were not better off in 1992 than they had been in 1988.

Bush had made a political error in waiting until January, 1992, to outline an economic plan. In his State-of-the Union address, the President announced a stimulus package that included the following:[91]

(1) a 90-day freeze on the imposition of new regulations
(2) a reduction in IRS income tax withholding
(3) a $5000 income tax credit for persons buying a home before the end of 1992
(4) a $500 per child increase in personal income tax exemptions
(5) a reduction in capital gains
(6) a tax credit of $3750 to allow the poor to buy health insurance
(7) repeal of the luxury tax on boats and airplanes

When a Republican President submits such a plan to a Congress controlled by Democrats in a presidential election year, its fate is largely sealed. Predictably, Congress rejected the Bush plan and adopted one of its own, which the President vetoed. The "gridlock" produced another Clinton campaign issue.

Philosophically, Bush was in favor of free trade - as are most Presidents - including Clinton. Later, both Reagan and Bush were to become "allies" of Clinton in his battle to win congressional support for the North American Free Trade Agreement (NAFTA) in 1993. Bush also favored continued government efforts to privatize some government functions and to reduce social services.

If the President had a problem with economic-policy making, he was inclined to refer to it as "that vision thing." His opponents often charged that he lacked "vision" and a sense of purpose in guiding the nation's economy. He was less able than Reagan, who excelled at it, to project an "image" of an optimistic, symbolic leader President.

The Richmond, Va., Presidential Debate of 1992

An incident that occurred during the second 1992 presidential debate in Richmond, Va., illustrates Bush's problem in clearly articulating what his policies meant to the common man and woman in America.

In a somewhat unusual format, it was decided that questioners in the debate would be selected randomly from undecided voters in the Richmond area. One Marisa Hall, a 25-year-old woman, asked the President, Governor Clinton and Ross Perot:[92]

> "How has the national debt personally affected each of your lives? And if it hasn't, how can you honestly find a cure for the economic problems of the common people if you have no experience in what's ailing them?"

Bush, in retrospect, was hurt most by his response. He struggled to answer the question but never seemed to satisfy Ms. Hall.[93]

"Obviously it has a lot to do with interest rates...Are you suggesting that if somebody has means, then the national debt doesn't affect them?"

Ms. Hall, in the national television spotlight for the first time in her life, responded:[94]

"Well, I've had friends that have been laid off from jobs...I know people who cannot afford to pay the mortgage on their homes, their car payment. I have personal problems with the national debt. But how has it affected you, and if you have no experience in it, how can you help us, if you don't know what we're feeling?"

Many Republicans, watching the debate in the comfort of their living rooms, may have experienced a sinking feeling in their stomachs. Bush had fumbled the ball and Clinton was about to pounce on it. One of the more famous (or infamous, depending on your viewpoint) TV images of that night was the President looking at his watch. Time was running out on his campaign.

Clinton responded that as governor of a small state, he personally knew people hurt by economic conditions. He ascribed the mushrooming national debt to "12 years of trickle-down economics" and to "the grip of a failed economic theory (supply-side economics)." Clinton said that as President, the chief doctor of the economy, he would prescribe a program of more investment in jobs, education and control of rising health care costs.

For the record, Perot responded that the seriousness of the nation's economic problems had caused him "to disrupt my private life and business to get involved in this...That's how much I care about it." Perot also mentioned his upbringing in a family of modest means and his good luck as an adult. This had led him to believe, he concluded, that he owed it to the country's children as well as his own children to do something about the sick economy.

Unemployment continued to grow while the economy slumped during the Bush years. In 1992, about 9.4 million Americans (about 7.4 % of the labor force) were unemployed. The anticipated increase in the rate of savings never occurred. Ardent Liberal redistributionists loudly proclaimed that the income gap between wealthy and poor Americans was constantly and dramatically growing.

The National Bureau of Economic Research said that the recession began officially in July, 1990. Bad economic news continued for the President and the nation in 1991 and 1992. With bad "pocketbook" news, Bush began to slide in the polls.

After the Persian Gulf War, Bush was the most popular President in the history of American public opinion polls. Some political pundits wondered whether any Democrat would be willing to run in 1992 and many "leading" Democrats, like New York Governor Mario Cuomo and House Majority Leader Richard Gephardt of Missouri, decided not to do so.

The Bush campaign focused on foreign policy and national security issues. His ads strongly suggested that the governor of Arkansas lacked the experience and character to be trusted in the White House, while the President already had demonstrated his ability to come through in a crisis,

By 1992, however, the electorate considered the Persian Gulf and the Cold War things of the past. They were primarily interested in domestic issues. The Bush campaign had forgotten an old saying about voter attitudes: "What have you done for me lately?"

After the Democratic National Convention, the President was 20 percentage points behind Clinton in the public opinion polls. He could not close the gap.

XXXX. THE CLINTON ADMINISTRATION

The election of 1992 was dominated by economic issues. Ironically, the success of Presidents Reagan and Bush in bringing the Cold War to a successful conclusion had shifted the focus of American political debate away from foreign policy and national security issues to bread and butter concerns. That also was the focus of Clinton's successful 1996 campaign.

Public opinion polls indicated that the American public held Bush responsible for the recession of 1991-92. Although the President insisted that there was nothing fundamentally wrong with the U.S. economy and that recovery had, in fact, already begun, Clinton disagreed. The former governor of Arkansas charged that Bush and his Republican administration had neglected the economy. As a matter of campaign strategy, Clinton had a clear-cut focus. A sign, hung on the wall of Clinton's national campaign headquarters, stated, *"It's the Economy, Stupid."*

Clinton promised that he would end the recession, ensure long-range prosperity and make improvements in the nation's physical infrastructure, badly in need of replacement and repair. There would, of course, be significant government investment in "human capital." The middle class, he promised, would get a tax cut and the rich would pay their "fair share."

During the campaign of 1992, Clinton ran as a "New Kind of Democrat." The implication was clear enough: He was not the same kind of Liberal as George McGovern (1972), Walter Mondale (1984) or Michael Dukakis (1988), all of whom had headed the Democratic Party ticket in years of electoral disaster. Clinton talked about the necessity of getting the Federal budget under control and reducing the bloated national budget.

This kind of rhetoric was indeed new for a Democratic presidential nominee. The Arkansas governor sounded at times like he was running as a Republican. The independent, self-financed presidential candidacy of Perot gave the campaign of 1992 a somewhat unusual dimension.

Perot constantly talked about budgets and deficits, complete with pie-charts and graphs in television infomercials. Clinton, in essence, had an "accomplice" in Perot, a man who forced the incumbent President to talk about economics.

President Bush's real strength was in national security and foreign policy. Perhaps no Chief Executive in recent U.S. History has had a more impressive resume in these areas. But he was vulnerable, fatally vulnerable, in the area of economic policy.

The "character issue" did not appear to hurt Clinton either. Questions about alleged marital infidelities, avoiding the draft during the Vietnam War era, leading demonstrations against his own country abroad while a Fulbright Scholar and smoking marijuana all were raised. But none of them did much damage. American voters were focused on the economy. And clearly they were dissatisfied with its condition.

Clinton was elected President, but without any overwhelming mandate. He won only about 43% of the vote. President Bush got 37%, Ross Perot 19%. Presidents are chosen, of course, not by direct popular vote, but rather by the Electoral College, which often distorts the popular vote. Perot won zero electoral votes. Clinton got 357, Bush only 168.

To some extent, Clinton had failed to provide voters with very specific proposals for deficit reduction out of fear of antagonizing them.

As one study noted, "...while concerned about ending deficits, the voters were even more concerned with preventing the spending cuts and tax increases necessary to accomplish that goal."[95]

As a former governor of Arkansas, Clinton was well aware of the impact of Federal economic policies on the states. After the November, 1994, mid-term congressional elections in which a political "earthquake" took place and Republicans gained control of both houses of Congress, Clinton spoke of a "New Covenant." The President's State of the Union address in 1995 sounded to Liberals too much like Reagan's "New Federalism." The new Republican majority listened with interest, if skepticism as well.

But Clinton wanted to focus on avoiding costly duplication of programs and to accompany shifts of functions with more national government support of states and local communities. He also wanted Congress to reduce the burden that Federal requirements impose on some state and local programs.

Two years earlier, on February 17, 1993, Clinton had unveiled his economic program to the Democratic Congress. It contained three major parts: (1) deficit reduction, (2) investment in research and development, education and training, physical infrastructure and (3) short-term spending to stimulate the economy.

During his campaign, Clinton had urged the American voters to "end gridlock" by electing a Democratic President to work with a Democratic Congress. There need be no more of the kind of conflict that characterized relations between Republican Presidents (Reagan and Bush) and Democratic Congresses. He may have been surprised by what happened to his first budget, submitted to a "friendly" Democratic-controlled Congress.

A titanic struggle occurred when the new President submitted a budget deficit-reduction plan involving a combination of spending cuts and tax increases. It took a gruelling summer of negotiations with congressional leaders, many of them Democrats, before the 1993 budget bill finally was approved.

The President originally had recommended a broad-based $72 billion energy tax. This was unsatisfactory to senators and representatives from oil and natural gas-producing states, particularly David Boren (D., Oklahoma) who refused to vote for the Clinton budget until the objectionable feature was deleted.[96]

Republicans argued that the new President was placing too much stress on the old Liberal Democratic idea of "tax and spend." Not enough attention was being paid to curtailing Federal spending, they argued.

Liberal Democrats were unhappy, too. The new President was not being sympathetic, they charged, to the plight of the poor and minorities. They were deeply concerned about domestic spending caps which Republicans and Conservative Democrats had forced upon Clinton.

Congressional Republicans were virtually unanimous in their opposition to the Clinton budget. So, too, were many Conservative Democrats. The Clinton plan passed the Senate, 51-50, only because Vice President Al Gore exercised his constitutional prerogative as Vice President (President of the Senate) to cast the tie breaking vote.

Gore would not have had the opportunity to break the deadlock, had Sen. Bob Kerrey not changed his mind. Kerrey was one of Clinton's rivals in the 1992 Democratic presidential primaries. He was relatively young, had served in Vietnam, lost his right leg below the knee from a grenade and won the Congressional Medal of Honor.

As a one-term bachelor governor of Nebraska, he had carried on a public relationship with Hollywood actress Debra Winger. When elected to the Senate in 1988, he became the first Medal of Honor winner to serve there since then Civil War.

Kerrey, however, failed to survive the primaries. He constantly hammered away at Clinton's avoidance of the military service during the Vietnam War. As a result, according to Washington Post editor Bob Woodward, he so alienated Hillary Clinton that she vetoed Kerrey's selection as her husband's vice presidential running mate.

Clinton, however, needed Kerrey's vote in 1993. Kerrey had called Clinton at the White House and announced his opposition to the Clinton budget. Woodward says in his book, *The Agenda: Inside The Clinton White House:*

"Clinton again pleaded with Kerrey that he needed his vote. 'My presidency's going to go down,' he said sharply, by now shouting. 'I do not like the argument that I'm bringing the presidency down!' Kerrey shouted back, getting fed up 'I really resent the argument that somehow I'm responsible for your presidency surviving,' Kerrey bellowed."

That particular call, according to Woodward, ended with an exchange of expletives. Ultimately Kerrey changed his mind. Rising on the Senate floor, he said:

"President Clinton, if you're watching now, as I suspect you are, I tell you this: I could not and should not cast a vote that brings down your presidency...Get back on the high road, Mr. President...Our fiscal problems exist because of rapid, uncontrolled growth in the (entitlement) programs that benefit the middle class...I'm sympathetic, Mr. President, I know how loud our individual threats can be. But I implore you, Mr. President, say no to us."

In the House, the Clinton plan passed, 218-216. Not one Republican voted for the bill, marking the first time since 1945 that the majority party in Congress had passed major legislation without any support at all from the minority party. *A switch of one vote in either House or Senate and Clinton's budget would have been defeated.*

One of the more interesting human stories of the House budget battle involved freshman Democrat Marjorie Margolies-Mezvinsky of Pennsylvania, the first Democrat to represent her suburban Philadelphia district in 76 years.

Margolies-Mezvinsky had promised during her campaign that she would not vote for any tax increases. But when she arrived in Congress, she promised that she would support the Clinton plan if her vote was necessary to pass it. She apparently did not believe that her vote would be decisive. As she switched her vote from nay to aye, Republicans chanted "Goodbye Marjorie." They were right. In 1994, she was defeated in her re-election bid.[97]

In order to get its budget adopted, the Clinton Administration had to make many deals, some of them nearly humiliating.

What were the key provisions of the agreement?

Wealthier Americans would be moved into a new, higher tax bracket. The previous top rate was increased from 31% to 39%.

Upper income retirees would have to pay Federal income taxes on 85% of their Social Security benefits, rather than the previous rate of 50%.

About 25% of all spending cuts were to come from Medicare.

About $21 billion was added to expand the earned income tax credit, which goes to low income, working families.

A 4.3 cent increase in the Federal gasoline tax would occur. This made the new rate 17.4 cents per gallon.

Perhaps the most important feature of the 1993 Clinton budget was the agreement with Congress to put caps on appropriations that require specific cuts to finance new spending. The bill also capped discretionary spending at $539 billion in fiscal 1994 and allowed spending to increase only slowly by 1998 to $549 billion.

The new President learned some hard lessons in the politics of the budgetary process in his first year in office. He learned about congressional egos. He learned that wearing the Democratic Party label was no guarantee that a member of Congress would support a Democratic President.

The Republican Victory of 1994

The new Republican Congress, elected in 1994, had a very different agenda than did Clinton. Newt Gingrich of Georgia, the new House Speaker, had developed a so-called "Contract With America" during the campaign. Clinton promptly called it a "Contract On America," setting the stage for partisan conflict with the presidential election looming ahead.

What was the "Contract With America?" Perhaps more than anything else, it was a strategic political move. Nearly all Republican candidates for the House of Representatives signed a document in front of the Capitol. The document did not apply to GOP Senate candidates.

It promised, in part, to enact:

(1) The Fiscal Responsibility Act[98]

A balanced budget/tax limitation amendment and a legislative line-item veto to restore fiscal responsibility to an out-of-control Congress, requiring them to live under the same budget constraints as families and businesses.

(2) The Taking Back Our Streets Act

An anti-crime package including stronger truth-in-sentencing, "good faith" exclusionary rule exemptions, effective death penalty provisions, and cuts in social spending from this summer's crime bill to fund prison construction and additional law enforcement to keep people secure in their neighborhoods and kids safe in their schools.

(3) The Personal Responsibility Act

Discourage illegitimacy and teen pregnancy by prohibiting welfare to minor mothers and denying increased AFDC for additional children while on welfare, cutting spending for welfare programs, and enacting a tough two-years-and-out provision with work requirements to promote individual responsibility.

(4) The Family Reinforcement Act

Child support enforcement, tax incentives for adoption, strengthening rights of parents in their children's education, stronger child pornography laws and an elderly dependent care tax credit to reinforce the central role of families in American society.

(5) The American Dream Restoration Act

A $500 per child tax credit, begin repeal of the marriage tax penalty and creation of American Dream Savings Accounts to provide middle-class tax relief.

(6) The National Security Restoration Act

No U.S. troops under U.N. command and restoration of the essential parts of our national security funding to strengthen our national defense and maintain our credibility around the world.

(7) The Senior Citizens Fairness Act

Raise the Social Security earnings limit which currently forces seniors out of the work force, repeal the 1993 tax hikes on Social Security benefits and provide tax incentives for private long-term care insurance to let older Americans keep more of what they have earned over the years.

(8) The Job Creation and Wage Enhancement Act

Small business incentives, capital gains cuts and indexation, neutral cost recovery, risk assessment/cost-benefit analysis, strengthening the Regulatory Flexibility Act and unfunded mandate reform to create jobs and raise workers wages.

(9) The Common Sense Legal Reform Act
"Loser pays" laws, reasonable limits on punitive damages and reform of product liability laws to stem the endless tide of litigation.
(10) The Citizen Legislature Act
A first-ever vote on term limits to replace career politicians with citizen legislators.

The document concluded:

"Further, we will instruct the House Budget Committee to report to the floor and we will work to enact additional budget savings, beyond the budget cuts specifically included in the legislation described above, to ensure that the Federal budget deficit will be less than it would have been without the enactment of these bills. Respecting the judgment of our fellow citizens as we seek their mandate for reform, we hereby pledge our names to this Contract with America."

The 104th Congress seized the initiative from the President. It passed, but Clinton vetoed, a balanced budget bill which substantially reduced taxes, cut entitlement growth spending, including Medicare and Medicaid and welfare. The Republican-controlled Congress removed entitlement status from Medicaid and Welfare and funded them through block grants to the states. It sharply cut several popular discretionary programs. Clinton again wielded the veto.

It was not until January, 1996, that Clinton and the Congress agreed that the budget should be balanced by 2002. They were, of course, not agreed on how to do it.

The 1995-96 congressional session was characterized by several "government shutdowns" of non-essential services and constant wrangling between Clinton and the GOP-controlled Congress, led in the House by Speaker Gingrich and in the Senate by Bob Dole, the party's presumptive 1996 presidential nominee.

Dole decided in May, 1996, to resign from the Senate and devote all of his energy to the campaign. At that time, he was nearly 20 percentage points behind Clinton in the polls. Dole declared that he was willing to put his fate in the hands of the American voters. He was either going to the White House or back to Kansas.

The electorate gave President Clinton 49% of the vote, Senator Dole 41% and Ross Perot 8%. The Electoral College vote was a lopsided 379 for Clinton, 159 for Dole. Perot got no electoral votes.

***

XXXXI. THE FUTURE OF AMERICAN CAPITALISM

"The free enterprise system is dying. We are gathered here around its deathbed. The initiative, the innovation, the hope and freedom that have given this country's people more of the fruits of life and liberty than mankind has ever enjoyed are dying with it....We should know that when it dies here, it will not rise again. And we should know that when it dies here, a long night of mediocrity will descend on us and our children."
-- Cornell C. Maier, Kaiser Aluminum & Chemical Corp. president, in a speech at a Miami University of Ohio business conference.

Few questions stir more heated debate among ideologue economists than, "Can capitalism survive?"

While many business leaders, economists and politicians see a bright future for America, others do not. Some professional economists, true to the tradition of the gloomy prophets of the past, insist that capitalism has a limited future, both in America and in the world.

Robert L. Heilbroner, Norman Thomas professor at the New School for Social Research, has grown progressively more pessimistic in recent years. Gone from his writing is the humor which marked *The Worldly Philosophers.* Two of his books, *An Inquiry Into the Human Prospect* and *Business Civilization in Decline,* are gloomy in tone. Such a perspective is traditional in economics, one of the reasons the discipline is called "The Dismal Science."

In *Business Civilization,* Heilbroner wrote:[99]

"I still believe that the civilization of business - the civilization to which we have given the name capitalism - is slated to disappear, probably not within our lifetimes, but in all likelihood within that of our grandchildren and great grandchildren."

Heilbroner's thesis is that the political apparatus within capitalism is steadily growing, enhancing its power, and usurping functions formerly delegated to the economic sphere.

"In the end," he writes, "I think this same political expansion will be a major factor in the extinction of the business civilization."

Heilbroner sees three kinds of problems, present in differing degrees in all capitalist economies. These are:

(1) The tendency of capitalism to develop generalized disorders that require government intervention, particularly problems of inflation and depression.

(2) The tendency to develop serious national disorders, such as the near breakdown of mass transportation in the U.S., the near collapse of the financial structure in Europe and the U.S. in the 1970s and the near insolvency of American and foreign cities.

(3) The dangers imposed by a constricting environment.

"There is," Heilbroner says, "a growing unease over the damage that unconstrained industrial expansion works on the life-carrying capabilities of the planet...Much has been written about these environmental challenges and their obvious implications with respect to the extension of the government's role within the economic system."

Heilbroner sees a drift toward planned capitalism. Without planning, he argues, capitalism cannot survive even in the short run.

But it is in the long run that capitalism will fail, Heilbroner says.

"...Not even the staunchest defender of capitalism...would assert that our business civilization will weather the next 1000 years or even the next 500."

Capitalism has been based on economic growth, Heilbroner says. Environmental restraints in the next century will constrict this expansive drive. The "spirit" of capitalism will be exhausted, he contends.

The inescapable increase in planning will extend the authority of government and replace private decision-making, he says. Planners will decide income distribution as well as what and how much is produced.

But what of individual freedom in the future? Can a fully-planned society be truly free? Heilbroner expressed some reservations about this:[100]

"The ideas of political equality and dissent, of intellectual adventure, of social nonconformity, however much hedged or breached in practice, owe much of their development to bourgeois thought...There are assuredly glories to the civilization of capitalism...The philosophy of individualism...like all deep beliefs, is an untestable proposition. Whether it will survive the demands of the future, I do not know...It at least offers the deepest reasons to hope that not all of its civilization will disappear with the business system."

Others are greatly concerned about future freedoms. In *Commentary* (April, 1978), Irving Kristol, editor and professor of Urban Values at New York University, wrote:

"...Whereas the older liberalism of the New Deal...aimed to help people lead the kinds of lives they wished to lead, the newer liberalism is far more interested prescribing - through bureaucratic directives - the kinds of lives they ought to live. This species of liberalism can only end up in the same place that more candidly socialist movements end up; a society where liberty is the property of the state and is or is not doled out to its citizens along with other contingent 'benefits.'"

Charles Frankel, professor of Philosophy and Public Affairs at Columbia University, wrote in *Commentary Symposium:*

"...a competitive economy and a relatively free market perform an invaluable service for a democratic government. Eliminate them, and the burden of deciding what a fair distribution of the social product entails falls entirely on the political system. He says that a government that undertakes this task will be subjected to pressures and grievances that will put its commitment to democratic modes of procedure under fearful strain."

In his Miami University address, Maier asked:

"Will we retain our individual rights and liberties as a people, or will we surrender them to an all pervasive state that will decide what's best for us...? Will we have a country where individual excellence is rewarded and achievement encouraged; or will we have a land where excellence and achievement are penalized?"

***

XXXXII. AMERICA'S POLITICAL ECONOMY: A RESUME

"...Nearly all people throughout history have been very poor. The exception, almost insignificant, in the whole span of human existence, has been the last few generations in the comparatively small corner of the world populated by Europeans. Here, and especially in the United States, there has been great and quite unprecedented affluence."
--John Kenneth Galbraith, *The Affluent Society*

Headlines often stress bad economic news. One has read of energy shortages, high unemployment rates and inflation. Taxpayer resentment and frustration, reflected in the dramatic case of Proposition 13 in California, raise serious questions about how large a role government should play in a free enterprise system.

Many middle-income families find it increasingly difficult to purchase homes and pay property taxes on them. Both steak on the table and nights out for recreation become rare in some families.

But all this brings one back to the fundamentals of economics. Most people, from the dawn of economic history, have been confronted with *scarcity*. They have not had enough resources to fill all of their needs and wants.

American resources, while abundant when compared to most other nations, are *finite*. We must recognize that no economic system can meet all of the needs of its people, particularly when *"need" tends in the U.S. to be defined as "wants."* We cannot have the proverbial cake and eat it too, however strongly we might wish to do so.

The American economy, like others, has established *priorities* among wants and needs. Our history of virtually continuous economic growth and development, and the boundless optimism produced as a result, may tend to obscure the fact that *resources are limited.* In short, some Americans expect too much from their economy.

Any system of *political economy* is a human invention. While American free enterprise has produced a standard of living that has made it the envy of the world, it is a poor cathedral in which to worship. As contemporary American capitalism bears comparatively little resemblance to that envisaged by Adam Smith in 1776, so American capitalism of the future may bear little resemblance to that of today.

One of the qualities that has enabled America to prosper has been our highly *pragmatic attitude*. Americans historically have been more interested in "what works?" than in particular mind sets produced by ideologies. If our past is any guide to our future, then Americans will evolve a system that fits their particular needs.

Economic growth is an important part of our legacy and it must continue if the size of the total economic pie is to grow. Government, which plays a vital if limited role in our free enterprise economy, must pursue those policies which contribute to growth. In some cases this may require difficult political choices, including the decision to defer some social goals that cannot be achieved without damaging the entire economic system.

Perhaps most important, if the American economy is going to continue to be "successful," it must win the support of the American people. This can perhaps best be done by economic education and by providing the basic necessities of life.

Public opinion polls indicate that:
- Americans want steady jobs and the opportunity to realize their human potential in their work.
- Americans want government to fight inflation so that their gains in income will be real, rather than illusory.
- Americans want adequate food, clothing, shelter, education and medical care.

These wants present a real challenge to the American *political economy*. Leaders of government, labor and industry have worked together to try to meet these challenges in the past. They have sought, in the past three generations, to avoid the kinds of destructive swings of boom and bust - such as the Great Depression: high unemployment and dangerous inflation.

Despite gloomy predictions from some economists that we shall witness the death of free enterprise in the 21st Century, there is some reason for hope, optimism and faith in the future of American capitalism.

While our contemporary political economy is certainly more complex than it was 200 years ago, we tend to forget or underrate the enormous growth of our human resources - the American people. We are more than 260 million strong, comparatively well educated and, some would say, more democratic and more open as a society than ever before.

It seems reasonable to assume that, given our resources, both human and material, the U.S. should be able to achieve its paramount national goals, once agreement on those goals has been reached through the democratic process. It also seems entirely possible, if not clearly probable, that Americans will reduce unemployment to socially acceptable levels, will evolve an energy policy that fits the national self-interest and achieve still further progress in education, health care delivery and social welfare, all within the framework of the free enterprise system.

American capitalism has many vocal critics. The system has been held up to the spotlight of much public soul-searching about its shortcomings. Perhaps ironically, this soul-searching has occurred precisely when the failures of alternative economic systems have been clearly revealed.

The now defunct Soviet Union had to import grain from the United States and Canada and bid for some kinds of technology from the West, particularly the United States. The failed authoritarian socialist economic system of the U.S.S.R. and its Eastern European satellites seem unlikely models to deliver a more abundant economic life than free enterprise.

Arthur Okum, chairman of the Council of Economic Advisers during the Johnson Administration, said:

"A market economy helps safeguard political rights against encroachment by the state. Private ownership and decision-making circumscribe the power of government or, more accurately, of those who run the government - and hence its ability to infringe on the domain of individual rights."

REFERENCES

1. Paul Sweezy, "Capitalism for Worse," in Leonard Silk (editor), *Capitalism: The Moving Target* (New York: Quadrange Books), pp. 121-128

2. Studs Terkel, "Here Am I, A Worker," in *Ibid.*, p. 68

3. *The Cincinnati Enquirer,* September 25, 1977

4. Charles W. Kegley Jr. and Eugene R. Wittkopf, *World Politics: Trend and Transformation* (New York: St. Martin's Press, 1995), p. 123

5. *Ibid.*, p. 125

6. *United Nations Demographic Yearbook, 1993*

7. Joshua S. Goldstein, *International Relations,* second edition (New York: Harper Collins, 1996), p. 467 (See chapter 12, "The North-South Gap")

8. Milton Friedman, *Capitalism and Freedom* (Chicago: University of Chicago Press, 1962), pp. 7-21

9. See Alan Ebenstein, William Ebenstein and Edin Fogleman, "Socialism" in *Today's Isms,* 10th edition (Englewood Cliffs, N.J.: Prentice-Hall, 1994), pp. 1-38

10. For a good brief discussion of the core of Communist thought, see "Communism" in *Ibid.*, pp. 110-199.

11. Adam Smith, *The Wealth of Nations* (Indianapolis, Ind.: The Bobbs-Merrill Company, 1961), pp. 52-62

12. See chapter I (The Relation Between Economic Freedom and Political Freedom) in Friedman, *op. cit.*, pp. 7-21

13. *Ibid.*, p. 8

14. Wasily Leontieff, "Sails and Rudders, Ship of State," in Silk, *op. cit.*, pp. 101-104

15. Russell H. Conwell, "Acres of Diamonds" in Alpheus T. Mason, *Free Government in the Making*, third edition (New York: Oxford University Press, 1965), p. 597

16. *Ibid.*, p. 579

17. Zbignew Brezinski, *Between Two Ages: America's Role in the Technetronic Era* (New York: The Viking Press, 1970)

18. Fredrick Jackson Turner, *The Frontier in American History* (New York: Henry Holt and Company, 1920)

19. Hamilton's Report on Manufacturers may be found in Mason, *op. cit.,* pp. 341-348

20. Stuart Holbrook, *The Age of the Moguls* (Garden City, N.Y.: Doubleday and Company, 1945), pp. 79-80

21. Andrew Carnegie, "Wealth" in *North American Review,* June, 1889, vol. 148, pp. 653-664

22. *Ibid.*

23. William Graham Sumner, "What Social Classes Owe To Each Other," in Mason, *op. cit.*, pp. 583-590.

24. *Supra.,* Note 21

25. *Theodore Roosevelt: An Autobiography* (New York: Da Capo Press, Inc, 1985), pp. 590-598

26. *Dartmouth College v. Woodward,* 4 Wheaton 518 (1819)

27. John Kenneth Galbraith, *American Capitalism: The Concept of Countervailing Power* (Boston: Houghton Mifflin Company, 1956)

28. John Kenneth Galbraith, *The New Industrial State* (Boston: Houghton Mifflin Company, 1967), p. 76

29. Quoted in Frank Cormier and William J. Eaton, *Reuther* (Englewood Cliffs, N.J.: Prentice-Hall, 1970), p. 98

30. Foster Rhea Dulles. *Labor in America: A History*, fifth edition (Arlington Heights, Ill.: Harlan Davidson, Inc. 1991), pp. 182-3

31. *Ibid.,* pp. 183-184

32. *Ibid.*, pp. 188-189. The Supreme Court decision may be found in *Loewe v. Lawlor,* 208 U.S. 274 (1908)

33. Nelson M. Blake, *A Short History of American Life* (New York: McGraw-Hill, 1952), p. 446

34. Robert F. Kennedy, *The Enemy Within* (New York: Popular Library, 1960)

35. Associated Press report in *The Lansing State Journal*, May 17, 1996

36. *Ibid.*

37. Yale Brozen, *Concentration, Mergers and Public Policy* (New York: Macmillan Publishing Company, 1982)

38. Federal deregulation efforts began in earnest with President Ford. All Presidents since have pursued it with varying degrees of enthusiasm.

39. See James M. Burns, *Roosevelt: The Lion and the Fox* (New York: Harcourt Brace & Company, 1956)

40. Burton I. Kaufman, *The Presidency of James Earl Carter* (Lawrence, Kansas: University of Kansas Press, 1993) pp. 55-57

41. Robert L. Heilbroner and Aaron Singer, *The Economic Transformation of America*, third edition (Fort Worth, Tex., Harcourt Brace & Company, 1994), p. 277

42. Quoted in Robert L. Heilbroner, *The Making of Economic Society*, ninth edition (Englewood Cliffs, N.J.: Prentice-Hall, 1993), p. 136

43. *Ibid.*, pp. 139-141

44. *Edwards v. California*, 314 U.S. 160 (1941)

45. Heilbroner, *The Making of Economic Society, op. cit.*, pp. 143-144

46. Louis Koenig, *The Chief Executive*, fourth edition (Ft. Worth, Tex.: Harcourt Brace, 1996), p. 276

47. Edwin P. Hoyt, *The Tempering Years* (New York: Scribners, 1963), p. 336

48. *Ibid.*, pp. 66-81

49. Burns, *Roosevelt: The Lion and the Fox, op. cit.*, pp.234-241, "Roosevelt as a Conservative"

50. John Maynard Keynes, *General Theory of Employment, Interest and Money* (New York: Harcourt, Brace and Company, 1936)

51. Robert L. Heilbroner, *The Worldly Philosphers* (New York: Simon & Schuster, 1967), pp. 225-261

52. See, for example, Robert Eisner, "Our Real Deficits," *Journal of the American Planning Association*, vol. 57, no. 2 (Spring 1991), reprinted in Thomas R. Swartz and Frank J. Bonello, *Taking Sides: Clashing Views on Controversial Economic Issues,* seventh edition (Guilford, Conn.: The Dushkin Publishing Group, 1995), pp. 221-228

53. See the chapter on "Taft-Hartley" in James T. Patterson, *Mr. Republican: A Biography of Robert A. Taft* (Boston: Houghton-Mifflin, 1972), pp. 356-370

54. David McCulloch, *Truman* (New York: Simon & Schuster, 1992), pp. 468-471

55. John Molloy, "Senator Robert A. Taft Often Startled His Supporters" (two-part series), *The Cincinnati Enquirer,* August 6 and August 13, 1961

56. Dwight D. Eisenhower, *Waging Peace* (Garden City, N.Y.: Doubleday & Co., 1965), pp. 127-131.

57. See Sidney Kraus (editor), *The Great Debates* (Bloomington, Ind.: University of Indiana Press, 1962), pp. 348-368

58. *Ibid.*

59. John F. Kennedy, *The Burden and the Glory,* Alan Nevins, editor (New York: Harper & Row, 1964), pp. 204-206

60. *Ibid.,* pp. 221-225

61. For a good account of the West Virginia primary, see Theodore White, *The Making of the President, 1960* (New York: Atheneum Publishers, 1961), pp. 97-114

62. Lyndon B. Johnson, *The Vantage Point: Perspectives on the Presidency, 1963-1969* (New York: Holt, Rinehardt and Winston, 1971), p. 87

63. Rowland Evans and Robert D. Novak, *Nixon in the White House: The Frustration of Power* (New York: Random House, 1971), p. 182

64. Herman Kahn, *The Next 200 Years* (New York: William Morrow and Company, 1976), pp. 54-56.

65. Elbert V. Bowden, *Economics: The Science of Common Sense* (Cincinnati: South-Western Publishing Company, 1974), pp. 369-370

66. Anthony S. Campagna, *Economic Policy in the Carter Administration* (Westport, Conn.: Greenwood Press, 1995), pp. 40-47

67. White, *American in Search of Itself, op. cit.*, p. 417

68. Herbert Stein, *Presidential Economics* (Washington, D.C.: American Enterprise Institute, 1984), pp. 217-218

69. Campbell McConnell and Stanley Brue, *Economics,* 13th edition (New York: McGraw-Hill, 1996), p. 687

70. *Ibid.*

71. Bill Clinton and Al Gore, *Putting People First: How We Can All Change America* (New York: Times Books, 1992)

72. White, *America in Search of Itself, op. cit.,* p. 150-151

73. George Will, "Cut a Rising Regressive Tax, Unmask the Monster Deficit," *Los Angeles Times,* January 12, 1990

74. Carolyn Weaver, *Social Security's Looming Surplus* (Washington, D.C.: American Enterprise Institute, 1990), p. 6

75. Daniel P. Moynihan, *Came The Revolution: Argument in the Regan Era* (New York: Harcourt Brace Jovanovich, 1988), p. 132

76. David Stoesz, *Small Change: Domestic Policy Under The Clinton Presidency* (White Plains, N.Y.: Longman Publishers, 1996), p. 179

77. Campagna, *op. cit.,* pp. 47-57

78. Jack W. Germond and Jules Witcover, *Mad as Hell: Revolt at the Ballot Box, 1992* (New York: Warner Books, 1993), p. 271

79. "Does A Flat Tax Make Sense?" *Time,* January 29, 1996, pp. 22-31

80. Campagna, *op. cit.,* pp. 136-140

81. Friedman, *op. cit.,* p. 9

82. *Ibid.,* p.34

83. *Ibid.,* p. 133

84. White, *America in Search of Itself, op. cit.,* pp. 163

85. *Ibid.,* p. 405

86. *Ibid.,* p. 149

87. James W. Lindeen, *Governing America's Economy* (Englewood Cliffs, N.J.: Prentice-Hall, 1994), p. 76

88. McConnell and Brue, *op. cit.,* pp. 269-271

89. White, *American in Search of Itself, op. cit.,* p. 304

90. Michael Duffy and Dan Goodgame, *Marching in Place: The Status Quo Presidency of George Bush* (New York: Simon & Schuster, 1992), pp. 228-229

91. *Ibid.*

92. Germond and Witcover, *op. cit.*, p. 9

93. *Ibid.*

94. *Ibid.*, p. 10

95. John Frendreis and Raymond Tatalovich, *The Modern Presidency and Economic Policy* (Itasca, Ill.,1994), p. 314

96. Richard E. Cohen, *Changing Course in Washington: Clinton and the New Congress* (New York: Macmillan College Publishing Company, 1994), pp. 123-130

97. James Q. Wilson and John DiIulio, *American Government,* sixth edition (Lexington, Mass.: D.C. Heath and Company, 1995), p. 490

98. Newt Gingrich, Rep. Dick Armey and House Republicans, *Contract With America* (New York: Times Books, 1994)

99. Robert L. Heilbroner, *Business Civilization in Decline* (New York: W.W. Norton & Company, 1976), pp. 17-38

100. *Ibid.,* pp. 122-124

SELECTED REFERENCES

Allen, Frederick Lewis. *The Big Change*. New York: Harper and Brothers, 1952

Ball, Terence and Richard Dagger. *Political Ideologies and the Democratic Ideal.* Second edition. New York: Harper Collins, 1995

Birnbaum, Jeffrey H. and Alan S. Murray. *Showdown at Gucci Gulch: Lawmakers, Lobbyists and the Unlikely Triumph of Tax Reform.* New York: Random House, 1987

Blake, Nelson M. *A Short History of American Life*. New York: McGraw-Hill, 1952

Bowden, Elbert V. *Economics: The Science of Common Sense*. Cincinnati: South-Western Publishing Company, 1974

Brewster, Lawrence G. and Michael E. Brown. *The Public Agenda*. Third edition. New York: St. Martin's Press.

Brzezinski, Zbigniew, *Between Two Ages*. New York: Viking Press, 1970

Buckley, William F. Jr. *American Conservative Thought in the Twentieth Century*. Indianapolis, Ind.: The Bobbs-Merrill Company, Inc., 1970

Burns, James McGregor. *Roosevelt: The Lion and the Fox*. New York: Harcourt, Brace and Company, 1956

Campagna, Anthony S. *Economic Policy in the Carter Administration*. Westport, Conn., Greenwood Press, 1995

Carnegie, Andrew. *Autobiography*. Boston: Houghton Mifflin, 1920

Citro, Constance F. and Robert T. Michael, editors. *Measuring Poverty*. Washington, D.C.: National Academy Press, 1995

Clinton, Bill and Al Gore. *Putting People First: How We All Can Change America*. New York: Times Books, 1992

Cohen, Richard E. *Changing Course in Washington: Clinton and the New Congress*. New York: Macmillan College Publishing Company, 1994

Cormier, Frank and William J. Eaton. *Reuther*. Englewood Cliffs, N.J.: Prentice Hall, Inc., 1970

Derthick, Martha and Paul J. Quirk. *The Politics of Deregulation*. Washington, D.C.: The Brookings Institution, 1985

Duffy, Michael and Dan Goodgame. *Marching in Place: The Status Quo Presidency of George Bush*. New York: Simon & Schuster, 1992

Dulles, Foster Rhea and Melvyn Dubofsky. *Labor in America: A History.* Fifth edition. Arlington Heights, Ill.: Harlan Davidson, Inc., 1991

Dye, Thomas R. *Power & Society.* Seventh edition. Belmont, Calif.: Wadsworth Publishing Company, 1996

Ebenstein, Alan, William Ebenstein and Edwin Fogleman, *Today's Isms.* Tenth edition. Englewood Cliffs, N.J.: Prentice Hall, 1994

Eisenhower, Dwight D. *Mandate for Change.* Garden City, N.Y.: Doubleday & Company, 1963.

------ *Waging Peace: The White House Years, A Personal Account.* Garden City, N.Y.: Doubleday & Company, 1965

Eisner, Marc A. *The State in the American Political Economy.* Englewood Cliffs, N.J.: Prentice Hall, 1995

Finklestein, Joseph. *The American Economy: From the Great Crash to the Third Industrial Revolution.* Arlington Heights, Ill.: Harlan Davidson, 1992

Fine, Sidney. *Laissez-Faire and the General Welfare State.* Ann Arbor: University of Michigan Press, 1957

------ *Sit-down: The General Motors Strike of 1937-1937.* Ann Arbor: University of Michigan Press, 1969

Frendreis, John and Raymond Tatalovich. *The Modern Presidency and Economic Policy.* Itasca, Ill.: F.E. Peacock Publishers, Inc., 1994

Friedman, Milton. *Capitalism and Freedom.* Chicago: University of Chicago Press, 1962

Funderburk, Charles and Robert G. Thobaben. *Political Ideologies: Left, Center and Right.* Second edition. New York: Harper Collins, 1994

Galbraith, John Kenneth. *The Affluent Society.* Boston: Houghton Mifflin Company, 1958

------*The New Industrial State.* Boston: Houghton Mifflin Company, 1967

------*American Capitalism: The Concept of Countervailing Powers.* Boston: Houghton Mifflin, 1956

------*The Anatomy of Power.* Boston: Houghton Mifflin Company, 1983

------*Economic Development.* Boston: Houghton Mifflin Company, 1964

------*Economics & The Public Purpose.* Boston: Houghton Mifflin Company, 1973

Germond, Jack W. and Jules Witcover. *Mad As Hell: Revolt at the Ballot Box, 1992.* New York: Warner Books, Inc., 1993

Goldman, Peter, Thomas M. DeFrank, Mark Miller and Tom Mathews. *Quest for the Presidency, 1992.* College Station, Tex.: Texas A & M University Press

Goldstein, Joshua. *International Relations.* Second edition. New York: Harper Collins, 1996

Gompers, Samuel. *Seventy Years of Life and Labor.* (2 volumes) New York: E.P. Dutton & Company, 1925

Hamby, Alonzo L. *Beyond the New Deal: Harry S. Truman and American Liberalism.* New York: Columbia University Press, 1973

Heilbroner, Robert L. *The Making of Economic Society.* Ninth edition. Englewood Cliffs, N.J.: Prentice Hall, 1993

------*The Worldly Philosophers.* New York: Simon & Schuster, 1967

------*Business Civilization in Decline.* New York: W.W. Norton & Company, 1976

------*An Inquiry Into the Human Prospect.* New York: W.W. Norton & Company, 1980

------*21st Century Capitalism.* New York: W.W. Norton & Company, 1993

----and Peter Bernstein. *The Debt and the Deficit: False Alarms/Real Possibilities.* New York: W.W. Norton, 1989

------and Aaron Singer. *The Economic Transformation of America: 1600 to the Present.* Third edition. New York: Haracourt Brace, 1994

Helfgott, Roy B. *Labor Economics.* Second edition. New York: Random House, 1980

Hofstadter, Richard. *Social Darwinism in American Thought.* Boston: The Beacon Press, 1955

------*The American Political Tradition.* New York: Vintage Books, 1948

Holbrook, Stuart, *The Age of the Moguls.* Garden City, N.Y.:Doubleday and Company, 1945

Hoyt, Edwin P. *The Tempering Years.* New York: Charles Scribner's Sons, 1963

Hughes, John Emmet. *The Ordeal of Power: A Political Memoir of the Eisenhower Years.* New York: Atheneum, 1963

Johnson, Haynes. *In The Absence of Power: Governing America.* New York: Viking Press, 1980

Johnson, Lyndon B. *The Vantage Point: Perspectives on the Presidency, 1963-69.* New York: Holt,

Rinehart & Winston, 1971

Kahn, Herman. *The Next 200 Years.* New York: William Morrow and Company, Inc., 1976

Kegley, Charles W. and Eugene R. Wittkopf, *World Politics.* Fifth edition. New York: St. Martin's Press, 1995

Kennedy, John F. *The Burden and the Glory.* (Alan Nevins, editor) New York: Harper & Row, 1964

Kennedy, Robert F. *The Enemy Within.* New York: Popular Library, 1960.

Keynes, John Maynard. *The General Theory of Employment, Interest and Money.* New York: Harcourt, Brace and Company, Inc., 1936

Koenig, Louis. *The Chief Executive.* Sixth edition. Ft. Worth, Tex.: Harcourt Brace, 1996

Krugman, Paul. *Peddling Prosperity.* New York: Norton, 1994

Loucks, William N. *Comparative Economic Systems.* New York: Harper & Brothers Publishers, 1957

Mason, Alpheus T. *Free Government in the Making.* New York: Oxford University Press, 1965

------and Richard H. Leach. *In Quest of Freedom.* Englewood Cliffs, N.J.: Prentice Hall, Inc., 1959

Mackenzie, G. Calvin and Saranna Thornton. *Bucking the Deficit: Economic Policymaking in America.* Second edition. Boulder, Colorado: Westview Press, 1996

McConnell, Campbell and Stanley Brue. *Economics.* Thirteenth edition. New York: McGraw-Hill, 1996

McCulloch, David. *Truman.* New York: Simon & Schsuter, 1992

Meese, Edwin III. *With Reagan: The Inside Story.* Washington, D.C.: Regnery Gateway, 1992

Meiklejohn, Alexander. *What Does America Mean?* New York: W. W. Norton & Company, Inc., 1963

Moody, John. *Masters of Capital.* New Haven: Yale University Press, 1919

Moynihan, Daniel P. *Came The Revolution: Argument In The Reagan Era.* New York: Harcourt Brace Jovanovich, 1988

Okun, Arthur M. *The Political Economy of Prosperity.* Washington, D.C.: The Brookings Institution, 1970

Penny, Timothy J. and Steven F. Schier. *Payment Due: A Nation in Debt, A Generation in Trouble.* Boulder, Colorado: Westview Press, Inc, 1996

Perot, Ross. *United We Stand: How We Can Take Back Our Country.* New York: Hyperion, 1992

Peterson, Peter. *Facing Up: Paying Our Nation's Debt and Saving Our Children's Future.* New York: Simon & Schuster, 1994

Phillips, Cabell. *The Truman Presidency: The History of a Triumphant Succession.* New York: The Macmillan Company, 1966

Phillips, Kevin. *Boiling Point: Republicans, Democrats and the Decline of Middle Class Prosperity.* New York: Random House, 1993

Quaglieri, Philip L. *America's Labor Leaders.* Lexington, Mass.: D.C. Heath and Company, 1989

Theodore Roosevelt. *Theodore Roosevelt: An Autobiography.* New York: Charles Scribner's Sons, Inc., 1913

------*The New Nationalism.* New York: The Outlook Company, 1910

Samuels, Warren J. *The Classical Theory of Economic Policy.* New York: The World Publishing Company, 1966

Samuelson, Robert J. *The Good Life and Its Discontents.* New York: Random House, 1995

Schlesinger, Arthur M. Jr. *A Thousand Days: John F. Kennedy in the White House.* Boston: Houghton Mifflin & Company, 1965

------*The Crisis of the Old Order.* Boston: Houghton Mifflin Company, 1957

Silk, Leonard. *Economics in Plain English.* New York: Simon and Schuster, 1978

------ *Capitalism: The Moving Target* New York: Quadrangle Press, 1974

Smith, Adam. *The Wealth of Nations.* Indianapolis, Ind.: The Bobbs-Merrill Company, 1961 (the work originally was published in 1776)

Smith, Hedrick, *The Power Game: How Washington Works.* New York: Random House, 1988

Sorensen, Theodore C. *Kennedy.* New York: Harper & Row Publishers, 1965

Stein, Herbert. *Presidential Economics.* Washington, D.C.: American Enterprise Institute Press, 1994

Stoesz, David. *Small Change: Domestic Policy Under The Clinton Presidency.* White Plains, N.Y.: Longman Publishers, 1996

Strobel, Frederick R. *Upward Dreams, Downward Mobility: The Economic Decline of the American Middle Class.* Lanham, Md.: Rowman & Littlefield Publishers, Inc., 1993

Susser, Bernard. *Political Ideology in the Modern World.* Needham, Mass.: Allyn and Bacon, 1995

Swartz, Thomas R. and Frank J. Bonello. *Taking Sides: Clashing Views on Controversial Economic Issues.* Seventh edition. Guilford, Conn.: The Dushkin Publishing Group, Inc., 1995

Thomas, Norman C. and Joseph A. Pika. *The Politics of the Prsidency.* Washington, D.C.: Congressonal Quarterly, Inc., 1996.

Thurow, Lester. *The Future of Capitalism: How Today's Economic Forces Shape Tomorrow's World.* New York: William Morrow & Company, 1996

Truman, Harry S. *Year of Decision.* Garden City, N.Y.: Doubleday & Company, Inc. 1955

-----*Years Of Trial and Hope.* Garden City, N.Y.: Doubleday & Company, 1956

Turner, Frederick Jackson. *The Frontier in American History.* (New York: Henry Holt, 1920)

Weaver, Carolyn. *Social Security's Looming Surpluses.* Washington, D.C.: American Enterprise Institute

White, Theodore. *The Making of the President, 1960.* New York: Atheneum Publishers, 1961

------*The Making of the President, 1964.* New York: Atheneum Publishers, 1965

------*The Making of the President, 1968.* New York: Atheneum Publishers, 1969

------*The Making of the President, 1972.* New York: Atheneum Publishers, 1973

------*America in Search of Itself.* New York: Harper & Row, 1982

Winkleman, B.F. *John D. Rockefeller.* Chicago: John Winston Company, 1937

Woodward, Bob. *The Agenda.* New York: Simon & Schuster, 1994.